BEYOND EQUILIBRIA

Oxford Series in Ecology and Evolution

The Comparative Method in Evolutionary Biology
Paul H. Harvey and Mark D. Pagel

The Cause of Molecular Evolution
John H. Gillespie

Dunnock Behaviour and Social Evolution
N. B. Davies

Natural Selection: Domains, Levels, and Challenges
George C. Williams

Behaviour and Social Evolution of Wasps: The Communal Aggregation Hypothesis
Yosiaki Itô

Life History Invariants: Some Explorations of Symmetry in Evolutionary Ecology
Eric L. Charnov

Quantitative Ecology and the Brown Trout
J. M. Elliott

Sexual Selection and the Barn Swallow
Anders Pape Møller

Ecology and Evolution in Anoxic Worlds
Tom Fenchel and Bland J. Finlay

Anolis Lizards of the Caribbean: Ecology, Evolution, and Plate Tectonics
Jonathan Roughgarden

From Individual Behaviour to Population Ecology
William J. Sutherland

Evolution of Social Insect Colonies: Sex Allocation and Kin Selection
Ross H. Crozier and Pekka Pamilo

Biological Invasions: Theory and Practice
Nanako Shigesada and Kohkichi Kawasaki

Cooperation Among Animals: An Evolutionary Perspective
Lee Alan Dugatkin

Natural Hybridization and Evolution
Michael L. Arnold

The Evolution of Sibling Rivalry
Douglas W. Mock and Geoffrey A. Parker

Asymmetry, Developmental Stability, and Evolution
Anders Pape Møller and John P. Swaddle

Metapopulation Ecology
Ilkka Hanski

Dynamic State Variable Models in Ecology: Methods and Applications
Colin W. Clark and Marc Mangel

The Origin, Expansion, and Demise of Plant Species
Donald A. Levin

The Spatial and Temporal Dynamics of Host–Parasitoid Interactions
Michael P. Hassell

The Ecology of Adaptive Radiation
Dolph Schluter

Parasites and the Behavior of Animals
Janice Moore

Evolutionary Ecology of Birds
Peter Bennett and Ian Owens

The Role of Chromosomal Change in Plant Evolution
Donald A. Levin

Living in Groups
Jens Krause and Graeme D. Ruxton

Stochastic Population Dynamics in Ecology and Conservation
Russell Lande, Steiner Engen, and Bernt-Erik Sæther

The Structure and Dynamics of Geographic Ranges
Kevin J. Gaston

Animal Signals
John Maynard Smith and David Harper

Evolutionary Ecology: The Trinidadian Guppy
Anne E. Magurran

Infectious Diseases in Primates: Behavior, Ecology, and Evolution
Charles L. Nunn and Sonia Altizer

Computational Molecular Evolution
Ziheng Yang

The Evolution and Emergence of RNA Viruses
Edward C. Holmes

Aboveground–Belowground Linkages: Biotic Interactions, Ecosystem Processes, and Global Change
Richard D. Bardgett and David A. Wardle

Principles of Social Evolution
Andrew F. G. Bourke

Maximum Entropy and Ecology: A Theory of Abundance, Distribution, and Energetics
John Harte

Ecological Speciation
Patrik Nosil

Energetic Food Webs: An Analysis of Real and Model Ecosystems
John C. Moore and Peter C. de Ruiter

Evolutionary Biomechanics: Selection, Phylogeny, and Constraint
Graham K. Taylor and Adrian L. R. Thomas

Quantitative Ecology and Evolutionary Biology: Integrating Models with Data
Otso Ovaskainen, Henrik Johan de Knegt, and Maria del Mar Delgado

Mitonuclear Ecology
Geoffrey E. Hill

The Evolutionary Biology of Species
Timothy G. Barraclough

Game Theory in Biology: Concepts and Frontiers
John M. McNamara and Olof Leimar

Adaptation and the Brain
Susan D. Healy

Competition Theory in Ecology
Peter A. Abrams

Beyond Equilibria: Historical Approaches to Ecology and Evolution
Hamish G. Spencer

Beyond Equilibria

Historical Approaches to Ecology and Evolution

HAMISH G. SPENCER
Department of Zoology, University of Otago, Dunedin, New Zealand

Great Clarendon Street, Oxford, OX2 6DP,
United Kingdom

Oxford University Press is a department of the University of Oxford.
It furthers the University's objective of excellence in research, scholarship,
and education by publishing worldwide. Oxford is a registered trade mark of
Oxford University Press in the UK and in certain other countries

© Hamish G. Spencer 2025

The moral rights of the author have been asserted.

All rights reserved. No part of this publication may be reproduced, stored in a retrieval system,
transmitted, used for text and data mining, or used for training artificial intelligence, in any form or
by any means, without the prior permission in writing of Oxford University Press, or as expressly
permitted by law, by licence or under terms agreed with the appropriate reprographics rights
organization. Enquiries concerning reproduction outside the scope of the above should be sent
to the Rights Department, Oxford University Press, at the address above

You must not circulate this work in any other form
and you must impose this same condition on any acquirer.

Published in the United States of America by Oxford University Press
198 Madison Avenue, New York, NY 10016, United States of America

British Library Cataloguing in Publication Data
Data available

Library of Congress Control Number: 2024946831

ISBN 9780192858993

ISBN 9780192859006 (pbk.)

DOI: 10.1093/oso/9780192858993.001.0001

Printed and bound by
CPI Group (UK) Ltd, Croydon, CR0 4YY

Cover photograph: 'An Imperial Shag (*Leucocarbo atriceps*) coming in to land'
© Aleksandr Simonov/Shutterstock.com
Back cover photograph: 'Rock stack moving beyond the equilibrium'
© KEVIN ELSBY/Alamy Stock Photo

Links to third party websites are provided by Oxford in good faith and
for information only. Oxford disclaims any responsibility for the materials
contained in any third party website referenced in this work.

The manufacturer's authorised representative in the EU for product safety is
Oxford University Press España S.A. of El Parque Empresarial San Fernando
de Henares, Avenida de Castilla, 2 – 28830 Madrid (www.oup.es/en or product.
safety@oup.com).
OUP España S.A. also acts as importer into Spain of products
made by the manufacturer.

Acknowledgements

Many people have played an important role in helping me to develop the ideas in this book and I am grateful to them all in the different roles they have played. The first seeds of these ideas were sown many years ago when, as an undergraduate student at Auckland, I took a "reading paper" with Wayne Walker, who had a few years before convinced me of the possibility of becoming a mathematical biologist. As part of this course we read and discussed Bob May's monograph, *Stability and Complexity in Model Ecosystems*. The book was an eye-opener to me, showing how interesting biology and mathematics could be blended. The central issue in May's book was the counter-intuitive result that in mathematical models of ecological communities, complexity did not seem to give rise to stability.

Steps towards the resolution of May's conundrum were being developed by Peter Taylor, whom I met three years later as a PhD student at Harvard. I discuss some of these ideas in Chapter 9. Peter's conclusions inspired my own research, addressing a parallel paradox in population genetics identified by Dick Lewontin, who was my doctoral supervisor. Dick's mentoring has left an indelible mark on me, and his formative influence will be discernible throughout this book.

My original work here was part of a research report I wrote for Marc Feldman's amazing theoretical population genetics course, which I was fortunate enough to take during a semester at Stanford. I later published this work in conjunction with Bill Marks, then at Villanova, who (as another graduate student, Greg Mayer, pointed out to me) had also recognized the potential of an historical approach.

After my move to Otago, two of my own PhD students, Meredith Trotter and Bastiaan Star, expanded the population-genetic modeling, bringing their own perspective to these questions, as did postdoctoral fellow Rick Stoffels. Several undergraduates, working as "summer students," Cully Mitchell, Hannah Grant, Kai Chiew and Cal Walter, also made important contributions. This work is discussed in Chapter 8.

Russell Gray and Martyn Kennedy, then also at Otago, introduced me to their critique of optimal foraging theory, which again led to a research project with an historical perspective. I use these collaborative findings as an exemplar of historical contingency in Chapter 1, returning to them in Chapter 3.

I have also benefited from the sage advice from several people in the academic-publishing industry. Sam Elworthy, then at Princeton University Press, whom I met at several Evolution conferences, encouraged me to test the waters for interest in the topic of this book by publishing a short abstract of the ideas. This test became my "covid-lockdown project," appearing in 2020 in *Quarterly Review of Biology*. Subsequently, Ian Sherman of Oxford University Press shepherded the proposal through the approval process, and Charles Bath also gave me useful feedback on a draft chapter. Subsequently, Giulia Lipparini has patiently taken on the tasks involved in getting through to publication.

Adrian Paterson (Lincoln University) has long encouraged me to write up my ideas about the importance of history and evolutionary predictability. Phil Lester (Victoria University of Wellington) read the whole manuscript and gave me invaluable feedback on how to make my message clearer. Of course, I remain responsible for any remaining obscurities.

Historian of biology Diane Paul (U. Mass., Boston), with whom I have collaborated for almost 40 years, has naturally argued all along for the importance of history, albeit usually in a different context. Nevertheless, one small project with her, concerning the time taken for equilibrium to be reached in a genetic model, provided an exemplar for history as approach in Chapter 7. At least as importantly, Diane has taught me how to write better: more concisely, more precisely and more directly.

Ken Miller and Crid Fraser kindly shared their artwork with me.

My family—Abby, Ned and David—have been amazingly encouraging, especially during the course of writing this book, but also over several decades more generally. Whatever accomplishments I have made would have been impossible without their supportive love.

Contents

List of Figures — xii

1 Introduction — 1

1.1 The Goal of this Book — 1
1.2 An Illustrative Example: Competitive Feeding Fish — 3
1.3 The Centrality of Equilibria in Ecological and Evolutionary Models — 6
1.4 The Flavors of History — 8
1.5 Concluding Remarks — 11

2 The Equilibrium — 14

2.1 Equilibria in Ecology and Evolution — 14
2.2 Stability of Equilibria — 18
2.3 Structural Stability — 22
2.4 Stochastic Versus Deterministic Models — 24
2.5 Multiple Equilibria — 25
2.6 The Importance of Different Outcomes — 26
2.7 Concluding Remarks — 28

3 Contingency — 29

3.1 The Flavor of Contingency — 29
3.2 Passerine Introductions in New Zealand — 30
3.3 Competitive Feeding Fish — 32
3.4 Wright's Shifting-Balance Theory — 35
3.5 Different Outcomes in Competitive Communities — 37
3.6 Contingency in Perspective — 39

4 Constraint — 41

4.1 History as Constraint — 41
4.2 The Minimal Gene Set for a Free-living Cell — 42
4.3 The Minimal Gene Set for an Endosymbiont — 45
4.4 A Metazoan Example — 47

4.5	Gain of Function in a Microbial Experiment	48
4.6	Adaptive Radiation in Scottish Sticklebacks	49
4.7	Constraint in Perspective	51

5 History as Template   54

5.1	Template	54
5.2	Evolution of Agonistic Behavior in Birds	54
5.3	Coevolution of Pigeons and Lice	56
5.4	Biogeography of Limpets in the Southern Ocean	58
5.5	Template in Perspective	63

6 Chance, Chaos and Capriciousness   67

6.1	Chance, Chaos and Capriciousness	67
6.2	Chance	68
6.3	Chaos	72
6.4	Capriciousness	75
6.5	Chance, Chaos and Capriciousness in Perspective	79

7 Approach and Turnover   81

7.1	Approach and Turnover	81
7.2	Approach	81
7.2.1	Mutation-selection Balance	82
7.2.2	Masking Tay-Sachs in NYC	84
7.2.3	Mixing Approach and Chance: Mutation-selection Balance in a Finite Population	85
7.3	Turnover	86
7.3.1	Neutral Theory of Molecular Evolution	86
7.3.2	Island Biogeography	90
7.4	Approach and Turnover in Perspective	92

8 Construction Part 1: Explaining Allelic Diversity   94

8.1	Construction	94
8.2	The Paradox of Variation	95
8.3	Equilibrium and the Parameter-space Problem	96
8.4	Constructing Polymorphisms	100
8.5	The Paradox Remains	103
8.6	Evolutionary Construction in Perspective	105

9 Construction Part 2: Model Ecosystems and Theoretical Ecology — 108

9.1 The Ecological Paradox — 108
9.2 Recognizing the Ecological Paradox — 111
9.3 The Development of Ecological Complexity — 114
9.4 Construction of Ecological Complexity — 115
9.5 Numerical Simulations Generating Ecological Complexity — 116
9.6 The Paradox Unsolved? — 118
9.7 Current Questions — 118
9.8 Priority Effects — 120
9.9 Ecological Construction in Perspective — 121

10 Concluding Remarks — 124

10.1 Why Bother with Flavors? — 124
    10.1.1 Contingency versus Construction — 125
10.2 Multiple Flavors — 126
10.3 The Historical Dimension — 129
10.4 A Roadmap for Researchers — 130

Index — 135

List of Figures

Fig. 1.1 An example of historical contingency: Each row shows one of the four different outcomes of the distribution of fish in a two-patch environment, and the respective ahistorical and historical likelihoods. The large fish consume twice as many resources as the small ones; the habitat on the left has twice as many resources as that on the right. The resources on each side match the consumption on each side in all four scenarios shown, solutions of the Ideal Competitive-Differences Distribution (ICDD). The Ideal Free Distribution (IFD) describes the situation when there are twice as many fish on the left as on the right; undermatching occurs when there are fewer fish at the richer site than expected under the IFD, and overmatching when there are more. An ahistorical approach to estimating the likelihood of the different distributions looks at the fraction of possible combinations of large and small fish in each scenario. This calculation implies the IFD-matching solution (the 8:4 ratio) is by far the most likely, because it affords more possible combinations. If, instead, we assemble the system over time, introducing individual fish randomly and allowing them to choose the richer resource contingent on the fish already there, simulations reveal that the undermatching 7:5 ratio is the most likely. See text for details. — 4

Fig. 1.2 The positive connotations of equilibria and "balance" suggest a reasonable and fair approach. Statue of Lady Justice blindfolded and holding a balance and a sword, outside Haarlem City Hall, Netherlands. — 7

Fig. 2.1 Equilibrium in a model of heterozygote advantage. AA homozygotes have a fitness of 0.8 compared to Aa heterozygotes; aa homozygotes have a relative fitness of 0.9. Top: Frequency of allele A over time (p) for four different initial values. Given the fitnesses in the underlying model, the system comes to rest at a value of $p = ⅓$ irrespective of its starting point. Bottom: Change in allele A's frequency (Δp) as a function of p. The equilibrium corresponds to $\Delta p = 0$ (the horizontal dashed line). Note that there are, in fact, three equilibria: $p = 0$ and $p = 1$ are unstable; only $p = ⅓$ is stable. — 16

Fig. 2.2 Expected heterozygosity at the steady state predicted by the neutral model of molecular evolution. N_e is the effective population size; v is the neutral mutation rate; the compound parameter $\theta = 4N_e v$, which increases linearly with both population size and mutation rate. — 17

List of Figures • xiii

Fig. 2.3 The trajectories of a predator-prey model exhibiting a stable limit cycle, plotted in state space. The x-axis shows the numbers of prey; the y-axis the numbers of predators. A system starting with 700 prey and 50 predators iterates with gradually more extreme population sizes, closer and closer to the orange line (the limit cycle). Systems initiated with values outside the limit cycle (e.g., 200 and 15, respectively) also approach it, but with gradually less extreme sizes. 17

Fig. 2.4 Different classes of stability displayed by a ball resting on the black surface. Small perturbations from either of the two locally stable equilibria allow the ball to return to that equilibrium, whereas any change in position from the neutrally stable equilibrium causes the ball to move further away (to either of the stable equilibria). Note that a larger move away from a locally stable equilibrium may mean the ball does not return, so neither is globally stable. The red locally stable equilibrium, however, is metastable, since a large push to the right will cause the ball to drop to the lower ("better") yellow equilibrium. A neutrally stable equilibrium would correspond to a flat interval on the black surface. 19

Fig. 2.5 Allele frequency (q) of a deleterious recessive with selection coefficient $s = 0.1$, starting at $q = 0.8$. The equilibrium value of q is zero. The red line shows the change under the deterministic model of Equation 2.1; the dotted blue lines show the results of 10 stochastic simulations with genetic drift, coming from the assumption of a finite population of size 500. 25

Fig. 2.6 Equilibrium in a model of heterozygote disadvantage. AA homozygotes have a fitness of 1.2 compared to Aa heterozygotes; aa homozygotes have a relative fitness of 1.1. Top: Frequency of allele A over time (p) for four different initial values. Given the fitnesses in the underlying model, the system comes to rest at a value of $p = 0$ or 1, depending on its starting point. Bottom: Change in allele A's frequency (Δp) as a function of p. The equilibrium corresponds to $\Delta p = 0$ (the horizontal dashed line). Note that there are, in fact, three equilibria: $p = 0$ and $p = 1$ are locally stable; whereas $p = ⅓$ is unstable. 27

Fig. 3.1 From left to right: Yellowhammer (*Emberiza citrinella*); cirl bunting (*E. cirlus*); ortolan bunting (*E. hortulana*); common reed bunting (*E. schoeniclus*). 31

Fig. 3.2 Adaptive landscapes. (a) Population mean fitness for the model of heterozygote advantage in Box 2.2. Note that the fitness peak occurs at the allele-frequency equilibrium of $p = ⅓$. (b) Population mean fitness for a two-locus, two-allele model of selection. Note that, for this particular model, the global peak occurs for a population fixed for the a and B alleles. A population with high frequencies of the alternative A and b alleles, however, will be driven by selection towards fixation of those alleles, which corresponds to a local fitness peak. An increase in adaptation for this latter population requires that the population somehow crosses the adaptive valley (the saddle) to the higher peak. (c) Wright's original diagram drew the adaptive landscape as a topographical map, where the axes represent frequencies of genotypic combinations, so that nearby points on the

	landscape are similar genetically. The dotted curves—analogous to contour lines—connect points of equal fitness, with peaks indicated by plus signs.	36
Fig. 3.3	The mean number of different final equilibria as a function of the initial number of species in the computer simulations of Gilpin and Case (1976). The bars indicate ±1 standard error. Redrawn with permission after Gilpin and Case (1976).	38
Fig. 4.1	In the upper figure, the ball represents the organism early in development. The path it takes as it rolls to the right (developing as it goes) involves certain "decisions" about the path to be taken, which constrain the subsequent developmental decisions. In the lower figure, a decision to take the right-hand path at the first choice has already been made, thus constraining the second decision, which is about to be made.	42
Fig. 4.2	The number of protein-coding genes and the genome size of *Buchnera* strains symbiotic with 39 aphid species from different parts of the Aphididae phylogeny. The clustering of points of the same color shows that strains of *Buchnera* commensal with more closely related aphids have similar genome sizes and numbers of protein-coding genes.	45
Fig. 4.3	Photo of the human commensal skin mite, *Demodex folliculorum*, showing its transparent, worm-like body and four pairs of legs.	47
Fig. 4.4	Protocol for Lenski's long-term evolution experiment (LTEE) using *Escherichia coli*.	49
Fig. 4.5	Adaptive radiation in the three-spined stickleback (*Gasterosteus aculeatus*), with the ancestral diadromous form in the center (with the red dot) and derived freshwater and brackish forms (purple dots).	50
Fig. 5.1	The evolutionary relationships of pelecaniform threat behaviors and their evolutionary derivatives, according to van Tets (1965).	55
Fig. 5.2	Phylogenies of the columbiform hosts (on left) compared with those of their wing (upper) and body (lower) lice (on right). Host-parasite associations are shown by the lines between the phylogenies.	58
Fig. 5.3	Map of the Southern Ocean, showing the Antarctic Circumpolar Current (ACC), as well as several of the subantarctic islands mentioned in the text. The Southern Ocean is generally considered to be the seas south of the Subtropical Front (STF). Antarctic and subantarctic waters meet at the Antarctic Polar Front (APF).	59
Fig. 5.4	The phylogeny of the Antarctic/subantarctic clade of the marine limpet genus *Nacella*. The red arrows indicate the estimated dates of divergence of the respective subclades. Note that each species is restricted to a single location, usually one island or island group, with closer locations having more closely related species. These patterns suggest that gene flow is very limited.	60
Fig. 5.5	*Siphonaria lateralis* (▲) and *S. fuegiensis* (△) on rocks (left) and living closely associated with the bull-kelp *Durvillaea antarctica*, Kerguelen Island (right).	61
Fig. 5.6	COI haplotype network for *Siphonaria lateralis* and *S. fuegiensis*. Note that the two species are both widely distributed around the Southern Ocean and that distant locations share haplotypes. These patterns suggest frequent, ongoing gene flow.	62

List of Figures

Fig. 6.1 The number of replicate *Drosophila melanogaster* populations (out of 107) with different numbers of bw^{75} alleles at selected generations in the experiment of Buri (1956). In the initial Generation 0, all populations comprised 16 heterozygotes and so, of the 32 genes present, 16 were bw^{75} alleles. Each generation consisted of a random sample of eight males and eight females, modeling genetic drift. By Generation 19, almost 30 populations had lost all bw^{75} alleles and a similar number fixed for this allele. Some populations remained polymorphic, however. 70

Fig. 6.2 Allele frequency over time in a simple model of genetic drift and selection against a recessive allele. Drift and selection interact, and the allele is lost faster in smaller populations. 72

Fig. 6.3 Iterations of Equation 6.1, starting at $x_0 = 0.3$, for increasing values of a. The four plots show a globally stable equilibrium, a two-cycle, a four-cycle and, finally, chaos as a increases. (Technically, there is chaos only for $x_t \in \left[\frac{a^2}{4}\left(1 - \frac{a}{4}\right), \frac{a}{4}\right]$.) 74

Fig. 6.4 The probability density function of a Cauchy distribution (brown) with location parameter 0 and scale parameter 2 compared to that of a normal distribution (blue) with mean 0 and variance 2π. The Cauchy looks rather like a normal with more density in the tails. 75

Fig. 6.5 The arithmetic sample mean of an increasingly larger sample from the Cauchy distribution (blue line) and the normal distribution (red line) of Fig. 6.4. The red line starts off close to zero (the normal's true mean) and iterates ever closer, whereas the blue line has no long-term trend, since the Cauchy distribution has no true mean. 76

Fig. 6.6 The allele frequency in a simple model of selection in which the selection coefficient (s) changes randomly each generation. The black line shows the allele-frequency trajectory for the same series of s values as the red line, but in the reverse order. 77

Fig. 6.7 The change in allele frequency in a simple model of selection for different values of the selection coefficient (s). Intermediate allele frequencies give larger changes, irrespective of s. 78

Fig. 7.1 Effect of population size on a recessive lethal for two mutation rates, $\mu = 10^{-5}$ and 10^{-6}. The mean frequency of this mutant is shown by the red and pink lines, plotted against the right-hand axis. In an infinite population, the expected frequency is given by $q = \sqrt{\mu}$, hence ~0.0032 and 0.0010, respectively. The frequency in finite populations will differ among populations: some, particularly small ones, will have no mutants. The expected proportion of populations lacking the mutant is shown by the blue lines, plotted against the left axis. 86

Fig. 7.2 Observed heterozygosity over time in two replicate simulated finite populations (red and blue) under the neutral hypothesis, starting at 0 (monomorphism). Expected heterozygosity (black line) starts at 0.0. In all cases, $N_e = 50{,}000$ and $\nu = 5 \times 10^{-6}$ and hence $H_E \approx 0.0909$. Note that the two observed trajectories are very different; in addition, neither is close to the expected trajectory. 87

xvi • List of Figures

Fig. 7.3 The rates of immigration (blue lines) and extinction (red and pink lines) depend on the number of species on an island. Where these rates are equal determines the number of species present on the island. So, for example, for islands near the mainland (dark blue immigration rate), a small island (red extinction rate) has fewer species than a large island (pink extinction rate). 91

Fig. 8.1 The proportion of random n-allele fitness sets leading to a stable, full polymorphism (i.e., stable equilibria with n alleles). Note the log scale of the y-axis. 98

Fig. 8.2 The number of alleles (n) over time, starting with $n = 1$ in a typical simulation run of Spencer and Marks (1988). The number of alleles very rapidly builds up to $n \geq 4$, values that would be exceedingly unlikely to maintain polymorphism if fitnesses had been chosen randomly. 100

Fig. 8.3 The number of constructionist simulations (out of 1000) with n alleles at Generation 10,000. 101

Fig. 8.4 The number of generalized-dominance constructionist simulations (out of 2000) with n_c common alleles (i.e., those at frequencies of 0.01 or greater) at Generation 10,000. 105

Fig. 9.1 A numerical example of population sizes governed by the Lotka-Volterra equations of Box 9.1 (with $a = \alpha = \beta = 1$ and $b = 3$). The unique equilibrium is given by $\hat{H} = 3$, $\hat{P} = 1$. The upper graph shows the population sizes over time of the prey (blue; $H(t)$) and predator (red; $P(t)$), given initial conditions $H(0) = 6$ and $P(0) = 2$. Note that the system oscillates without approaching (or moving away from) the equilibrium. The lower phase-plot graph shows how $H(t)$) and $P(t)$ co-vary over time, depending on the initial conditions (history as contingency). The orange curve corresponds to the upper graph. 110

Fig. 9.2 Interactions (edges) between species (nodes) in two scenarios of community assembly, (a) "Immigration" and (b) "Radiation," in the models of Maynard et al. (2018). The obvious differences are a spandrel of the assembly process, not a directly selected property. See text for details. 120

Fig. 10.1 The inferred phylogeny of the chromosome-3 inversions in *Drosophila persimilis* and *D. pseudoobscura*. Each of the names represents a particular chromosomal arrangement, with each arrow indicating the origin of a new arrangement via a single inversion event. Note that "Standard" is found in both species. There is just one "hole" in this phylogeny: "Hypothetical" bridges the two-inversion change between "Santa Cruz" and "Standard." 127

Fig. 10.2 The distribution of the number of alleles produced by 1000 replicate runs of (top) the constant-viability model of Spencer and Marks (1988) and (bottom) the frequency-dependent selection model of Trotter and Spencer (2008). The left-hand panels show these distributions at Generation 10,000 ("snapshot"); the right-hand panels show the distributions after each simulation is allowed to run to equilibrium in the absence of novel mutation. Note that the distributions are very similar for the models of constant-viability selection, with the greatest difference being the elimination of outcomes with larger numbers of alleles (> 10) in the equilibrium distribution. By contrast, in the models of frequency-dependent selection, there is a significant shift to the left in the equilibrium distribution compared to that of the snapshot.   128

Fig. 10.3   Bar charts showing the distribution of the mean number of alleles and common alleles in the constructionist simulations of Spencer and Walter (2024), without environmental deterioration (a) and with environmental deterioration (b).   131

1

Introduction

1.1 The Goal of this Book

The concept of biological equilibrium pervades every aspect of ecology and evolution, from the popular notion of the "balance of nature" to sophisticated mathematical analyses of critical points in the state space of scientific models. Indeed, the concept has long been considered central to the whole of biology (Conference on Concepts of Biology, 1958). Whether we are working with verbal or complicated mathematical models, finding and evaluating the properties of the equilibria greatly expands our understanding of the model and, presumably, the underlying biology. But overly focusing on the equilibrium and analyzing our biological systems exclusively in terms of this aspect of our models implies that the time dimension—the *history*, if you like—is of secondary importance.

Indeed, as I discuss below, some theoretical ecologists in the twentieth century went further, intentionally taking an ahistorical view of biology. The rationale for such an approach, as historian of science Sharon Kingsland (1985) has argued, was the desire of these theorists to be able to make general conclusions, beyond the particularities of their study system. History was seen as contingency, a series of unique events that afforded no general insights. Since generalizing—the distillation of the essence of a system into principles that apply elsewhere—is so highly prized in science, excluding history ostensibly improved the quality of the theorists' ecological science.

Perhaps less obviously, evolutionary biologists have also been wary of explicitly addressing some aspects of the historical dimension of their science. In his agenda-setting survey of evolutionary genetics, Richard Lewontin (1974: 269) held that "population genetics is an *equilibrium theory*" and noted that "Equilibria annihilate history." He went on to assert "Evolutionary geneticists are anxious to purge historical elements from their explanations" due not only to epistemological reasons (once at equilibrium, any trace of history may be impossible to trace), but also to a socio-political "preoccupation with stability."

Philosopher and historian Michael Ruse (1996) has claimed that evolutionists have long avoided any discussion of biological progress in their professional science, even as they were happy to include it in their semi-popular writing. But this care also excluded some science: Ernst Mayr, in his role as first editor of the Society for the Study of Evolution's journal *Evolution*, rejected several papers that so much as hinted at evolutionary progress, including any that indulged in "phylogenetic speculations" (Ruse, 1996: 446; see also Spencer, 1998). (Today's evolutionary biology is awash with such trees, however, which suggests that they are no longer seen as so imbued with progress as they once were. Progress and phylogenies have apparently gone their separate ways.) Instead, Mayr wanted papers about such topics as adaptation, paleontology, *Drosophila* chromosomes, hybridization, taxonomy, speciation and population genetics (Cain, 1994).

In this book I argue strongly against this purely ahistorical approach. Rather, I contend that adding an explicit consideration of history—looking beyond the equilibrium—provides a deeper understanding of almost every eco-evolutionary system. Contrary to the view of the ahistorical ecologists, history matters in all sorts of ways (although knowing about the equilibrium is clearly relevant, too). For example, we should be asking about the path we took to get to equilibrium, even assuming we are at such a balance point, since we might have ended up at a different equilibrium. Or it might be that understanding that history allows us to make novel predictions about our study system that cannot be derived from its current state.

I further argue that a consideration of these different questions reveals that history comes in a number of *flavors*, by which I mean various ways in which history can play a critical role in biological processes. Like the flavors of our food, these historical flavors may not always be distinct: they may be blended or overlapping and any number may be present. But distinguishing among historical flavors is of practical use in explicating ecological and evolutionary phenomena. Much of this book catalogs examples drawn from my own and others' published accounts of ecological and evolutionary studies, comparing and contrasting ahistorical analyses with historical approaches that illustrate different flavors. In all cases, I contend that incorporating both ahistorical and historical approaches is likely to lead to a far more complete, albeit nuanced, understanding of biology. Ultimately, I hope to inspire the reader to examine their own study systems through an historical lens, as well as taking the more traditional, equilibrium-centered, ahistorical view.

I should make two disclaimers here. First, I am not claiming that equilibria are inconsequential, or that modelers should not be interested in equilibria. Elucidating and characterizing the equilibria of models, even those beyond biology, is fundamentally important. Moreover, some of the insights gained by using the historical approach I advocate are only apparent if various properties of the equilibria are already known. It is when we think solely about equilibria that I believe we miss out; I am arguing for a dual approach.

Second, I am not the first to carry out studies incorporating an historical approach, nor am I unique in emphasizing the importance of history (as indeed, the citations above demonstrate; see also Estes & Vermeij, 2022). As will become clear when I discuss various examples of different flavors in subsequent chapters, there is much eco-evolutionary research that has used an historical approach. (See, especially, the discussion of "priority effects" in Chapter 9.) What is novel is that I believe almost all research in ecology and evolution would benefit from taking history seriously, and that distinguishing among the flavors of history helps us do so more insightfully.

Last, I should also point out that much of the book discusses models, usually mathematical models of biological systems. Many of the points I make extend to verbal models and even beyond, but often the cases that best exemplify the phenomena I want to illustrate are mathematical models. Perhaps this is because the process of constructing a mathematical model is, by necessity, explicit about the assumptions concerning the biological factors and processes being examined. Moreover, many of these models are what Servedio et al. (2014) call "proof-of-concept models," designed to test the validity of verbal hypotheses, rather than to interface directly with real-world data. This is not to say that proof-of-concept models need not be biologically realistic, but (like our verbal explanations) they necessarily omit what are thought to be extraneous details in the interests of mathematical (or verbal) tractability.

1.2 An Illustrative Example: Competitive Feeding Fish

To make these ideas more concrete, let me take an example from behavioral ecology. Consider an environment with a resource, say food, that is patchily distributed. Can we predict the distribution of consumers of this resource (usually animals) across these patches? Assuming that selection maximizes the fitness of the consumers, which in turns means maximizing the net resource gained, optimal foraging theory makes such predictions. Note that this wording almost disguises that we are using an ahistorical approach, but the solution is an optimum, the fitness-maximizing equilibrium.

One of the simplest applications of optimal foraging theory is the Ideal Free Distribution (IFD), which sees the foragers making an unconstrained ("free") choice about where to forage based on perfect ("ideal") information about the availability of resources (Fretwell & Lucas, 1970). Consider a two-patch environment in which food is continuously available at respective rates R_1 and R_2. The IFD then predicts that if the numbers of the animals at the two sites are N_1 and N_2, respectively, their ratio matches that of the food availability: $N_1/N_2 = R_1/R_2$. If there are fewer consumers than expected at the good site, we say that we have "undermatching;" too many is called "overmatching". Notice, again, that we have slipped into using ahistorical language and are focused solely on an equilibrium.

Fretwell and Lucas (1970) point out that by "free" they do not just mean that an animal is able to select either site without restrictions; freedom entails that patch choice is on an equal basis with the other foragers, who are identical in their foraging abilities. Of course, not all animals are equal (Orwell, 1945); some are better competitors than others, for example. Nevertheless, we can easily extend the IFD to avoid the free assumption and incorporate such differences. Parker and Sutherland (1986) and Houston and McNamara (1988) do just that and note that the same optimality logic implies that the sums of the competitive abilities at each site, respectively C_1 and C_2, should satisfy $C_1/C_2 = R_1/R_2$.

4 • Beyond Equilibria

This extension of the IFD is no longer "free:" Spencer et al. (1995) call it the Ideal Competitive-Differences Distribution (ICDD). Perhaps more importantly, there is often more than one way that the animals can distribute themselves under an ICDD. In other words, there are several combinations of N_1 and N_2 that lead to $C_1/C_2 = R_1/R_2$. Let us explore this aspect of optimal foraging theory with a numerical example.

Consider a situation with two feeding sites, one twice as productive as the other ($R_1/R_2 = 2$), and 12 fish consuming that food, six of which can obtain twice as much food as the remaining six. There are actually four solutions (i.e., values of N_1 and N_2 satisfying $C_1/C_2 = R_1/R_2$) to this ICDD problem (see Fig. 1.1). For example, having all six good competitors at the more productive site and all six poor competitors at

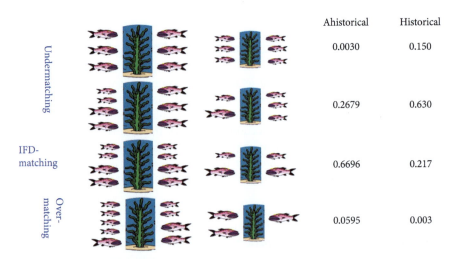

Fig. 1.1 An example of historical contingency: Each row shows one of the four different outcomes of the distribution of fish in a two-patch environment, and the respective ahistorical and historical likelihoods. The large fish consume twice as many resources as the small ones; the habitat on the left has twice as many resources as that on the right. The resources on each side match the consumption on each side in all four scenarios shown, solutions of the Ideal Competitive-Differences Distribution (ICDD). The Ideal Free Distribution (IFD) describes the situation when there are twice as many fish on the left as on the right; undermatching occurs when there are fewer fish at the richer site than expected under the IFD, and overmatching when there are more. An ahistorical approach to estimating the likelihood of the different distributions looks at the fraction of possible combinations of large and small fish in each scenario. This calculation implies the IFD-matching solution (the 8:4 ratio) is by far the most likely, because it affords more possible combinations. If, instead, we assemble the system over time, introducing individual fish randomly and allowing them to choose the richer resource contingent on the fish already there, simulations reveal that the undermatching 7:5 ratio is the most likely. See text for details.

After Spencer (2020), with permission of the University of Chicago Press. © University of Chicago Press (2020).

the poorer site leads, as required, to twice as much consumption at the better site. But so does having four of each competitive class at the better site and two of each at the poorer site. Notice that this latter scenario mimics the IFD, since $N_1/N_2 = R_1/R_2$.

An obvious question to ask is, "Which scenario is the one we are most likely to observe?" One way to answer this question is to evaluate the different numbers of combinations of good and poor competitors that give each of the solutions (Houston & McNamara, 1988). For example, the IFD-matching solution requires that four of the six good competitors are selected, as are four of the six poor ones. The number of ways we can make such a choice is given by the product of the binomial coefficients,

$$\binom{6}{4}\binom{6}{4} = 225.$$

If we do the parallel calculations for all four solutions shown in Fig. 1.1, we find, respectively, 1, 90, 225 and 20 combinations, a total of 336. The IFD-matching scenario is clearly the commonest, being selected $225/336 \approx 67\%$ of the time. Note that this ahistorical ICDD analysis focuses exclusively on the equilibria.

But what happens if we take history seriously and look beyond the equilibrium to ask, "Which scenario is the most likely to develop over time?" Computers allow us to explore an explicitly historical approach, running simulations in which consumers arrive in the system in a random order and make individual decisions on which site to choose, contingent on where the earlier-arriving consumers have already gone. Such an approach gives an entirely different set of outcomes (Spencer et al., 1995; see Fig. 1.1). Now, the most likely distribution is numerically undermatching: five good and two poor consumers (seven in total) at the better site with one good and four poor consumers (five in total) at the poorer.

This comparison might seem rather esoteric; why do we care about differences between the ahistorical and historical distributions? Both approaches allow all four solutions, after all. In short, it matters because, as Kennedy and Gray (1993) demonstrated, undermatching is a common, but often unnoticed, finding in both laboratory and field experiments. Incorporating history means we can explain this phenomenon; the ahistorical approach cannot. In the above example, for instance, if we simply count the number of fish at each site—perhaps because we are ignorant of any competitive differences, or we cannot easily measure them—undermatching is almost three times more likely (78% versus 27%) in the historical simulations than in the ahistorical combinations.

This example also shows how revealing it is to analyze the model both historically and ahistorically. It only becomes apparent how illuminating the former approach is when we realize that it predicts undermatching whereas the latter does not. The importance of being able to make this contrast will be a recurring feature of examples later on in the book.

Kennedy and Gray (1993) suggested a number of reasons for widespread undermatching, involving violations of both the ideal and free assumptions of the IFD. For instance, consumers might be unable to perceive small marginal differences in the benefits of foraging at one site versus the other (i.e., they had "perceptual limits"). If, in such circumstances, they chose a site randomly, the ideal assumption would not be satisfied. Computer simulation studies using an historical approach show that, indeed, perceptual limits lead to undermatching (Spencer et al., 1995). Similarly, as we saw above, competitive differences violate the IFD's free assumption, but, crucially, it is only with an historical approach that we see that they also lead to significant undermatching (Spencer et al., 1995).

So clearly history is important; acknowledging its effect gives us greater insights and explains real, unexpected observations. The way in which history affects this model system is the flavor I call *contingency* because the incoming fish choose their site (that which offers more resource when they arrive) contingent on the fish already in the system. There are also elements of the flavor of *construction*, which emphasizes how a system is constructed (or assembled) over time. The archetypal examples of history as construction, however, lead to a dynamic equilibrium (or steady state) (see Chapters 8 and 9).

Finally, we should note that, in this example, both historical and ahistorical approaches focus on the equilibrium, i.e., the final distribution of fish. Only the former, however, asks how we got there, and that turns out to make all the difference.

1.3 The Centrality of Equilibria in Ecological and Evolutionary Models

Ecologists and evolutionists make dynamical models for many reasons (Servedio et al., 2014). All such models are at some level verbal, but many are distilled into mathematical descriptions in the form of equations. Most dynamical models attempt to mimic the temporal behavior of the biological system by abstracting what the scientist thinks are its critical features. Such models can be used for the analysis and interpretation of observational or experimental data, as well as for predicting currently unknown or unknowable outcomes (e.g., long-term behavior or how the system might respond if perturbed in some way). Other models are set up as a kind of null hypothesis to be rejected, allowing a conclusion that the model is inadequate in some way and hence pointing to biological phenomena that need to be incorporated into a better model or assumptions that need to be modified. Whatever the case, the motivation is surely to gain insight into the system. Even the intellectual process of constructing the model can be informative, but normally the analysis of the model provides the greatest understanding. For most models that means mathematical analysis of one or more equations that capture the essence of the verbal description.

So how do we analyze dynamical models in ecology and evolution? Given the temporal dimension, it is only natural to ask when and where the system—or at least the model system—goes; does it come to rest or does it forever keep changing? In other words, are there equilibria and how can we characterize them? Fortunately, mathematics provides a number of tools to locate and characterize equilibria (see Box 2.1, for instance). Moreover, it is clear that knowing about a model's equilibria vastly improves our understanding of the behavior of the model, even if the biological system is not at an equilibrium. For example, mathematical analyses can tell us if the system is likely to approach an equilibrium and how quickly it might do so; this sort of conclusion might be important in understanding how robust our real-world biological system is to disruptions. More abstractly, we may be able to deduce general features of a model, such as the direction of change in the mean fitness of a population under selection, properties that may not be beholden to the particular values of the model's parameters.

So, in short, analyzing a model to discover its equilibria and their properties is a natural way to proceed: there are well-understood tools available and they provide valuable answers. Nevertheless, it seems to me that there are at least two further reasons for equilibria being the focus of model analysis in ecology and evolution.

First, there is the overwhelmingly positive connotation of the word *equilibrium*. Equilibrium means balance, composure, steadiness, equanimity, calmness; all of these words convey the idea that equilibrium is a desirable good. We conjure up visions of a blinded Lady Justice and the scales of justice on which she is dispassionately weighing up the conflicting arguments pertaining to a sensational court case (see Fig. 1.2). The morally correct decision comes from her balanced consideration of the opposing points. We are comforted by the view that things are fair.

Fig. 1.2 The positive connotations of equilibria and "balance" suggest a reasonable and fair approach. Statue of Lady Justice blindfolded and holding a balance and a sword, outside Haarlem City Hall, Netherlands.

Photo by Till Niermann, reproduced under the Creative Commons Attribution-Share Alike 3.0 Unported license.

There is no doubt that this positivity extends into the realm of ecology and evolution. Popular writing about ecological matters often talks about the balance of nature and how disturbing this balance leads to all sorts of problems. Kingsland (1985) notes that right from the beginning of scientific considerations of populations, equilibrium has been sought and found in nature. Charles Darwin argued that the struggle for existence leads to equilibrium: "in the long-run the forces are so nicely balanced, that the face of nature remains uniform for long periods of time" (Darwin, 1859: 73). The natural world contains a system of checks and balances that, amongst other things, prevents any species from increasing its number too greatly. One of the first ecologists to publish on mathematical models of density dependence in population size, A. J. Nicholson, was motivated in his work to understand why population sizes appeared, at least approximately, to be stable (Kingsland, 1985).

In evolutionary theory, another historian of science, John Beatty, has suggested that the bitter classical/balance controversy of 1950s and 1960s population genetics was named by one of the protagonists, Theodosius Dobzhansky, in order to give his "balance school" a modern, up-to-date cast compared to the old-fashioned "classical school" of his opponent, Hermann Muller. This slur was certainly recognized by Muller, who wrote to one of his collaborators in 1958, "How strange it seems to me to have things termed classical, with the implication that they are out of date" (Beatty, 1987).

Second, and probably more importantly, focusing on equilibria means we can excise history from our analyses. As I outline above, Kingsland argued that the mathematical ecologists of the 1960s eschewed history because they felt it smacked of particularities and allowed few generalizations (Kingsland, 1985). In the same way that natural historians are viewed as vestiges of nineteenth-century amateurs and not really professional scientists, history was seen as antithetical to genuine biological research, a narrative account that affords little scientific progress (Spencer, 2020).

Indeed, this disparaging attitude to historical accounts remains widespread: while ecologists and evolutionists pay lip service to the importance of biological history to their favorite study system, the inclination is often to omit it from the analysis of their models and explanations, writing it off as contingency, unpredictable by any law. The ahistorical view, by contrast, is highly privileged: "solving" a model often reduces to finding and characterizing the equilibria, and the focus rarely moves beyond these equilibrial properties of the system. But the above example from optimal foraging theory shows that these attitudes are wrong: general principles that apply outside the study system can be distilled using tools that explicitly acknowledge history.

1.4 The Flavors of History

In addition to arguing that history is crucial in understanding ecology and evolution, I contend that history can impact biological systems in a number of subtly different ways, what I call flavors. So what are these flavors? Much of the rest of this book expounds in detail just what I mean, giving concrete examples and showing how the flavors can blend and complement each other. For now, however, let me outline a brief schema.

Undoubtedly, the commonest depiction of the role of history in ecology and evolution is *contingency*: the occurrence of events is conditional on the properties of the system in question. In the historical studies of the foraging fish, for instance, each individual entering the system chose the marginally more profitable site, this decision being contingent on the previous choices of the fish that had already arrived. The ahistorical bias of Kingsland's ecological theorists develops from the argument that what happened in the past was contingent on so many actors and their actions that any generalizing would be impossible. But the fish show this conclusion is false: a mathematical model with contingency can lead to broad inferences that help us understand real-world observations.

A second flavor of history is *constraint*, in which the possible events are constrained by previous events. For instance, evolutionary biologists are interested in working out which genes are indispensable in the genome reduction that has occurred in eubacterial intracellular parasites (the "minimal gene set"). In many cases, an ahistorical approach looking at the genes present gave too small a list, whereas an historical approach, examining the evolution of different parasite lineages, revealed that the loss of a particular gene (say, Gene A) precludes the loss of a second (Gene B). Gene B was now essential, but it could have been successfully lost earlier if that loss had preceded the loss of Gene A. Constraint is stronger than contingency; constraint rules out certain possibilities, whereas contingency alters the relative likelihoods of certain events.

A third way of viewing history is as a *template* on which change occurs, such as the use of historical hypotheses (e.g., phylogenies) in eco-evolutionary research. Here we use an historical approach to explain particular evolutionary changes: unique events need not be beyond the reach of general explanatory methods. For example, Johnson and Clayton (2003) showed that the wing lice that parasitize pigeons and doves have done so for a much longer time than these birds' body lice. By necessity, explanations of unique historical events are particular to a single biological system (say, lice and pigeon coevolution); what is general is, of course, the method.

Three further flavors of history—*chance, chaos* and *capriciousness*—are closely related. Chance is evident in what appear to be random processes like genetic drift and mutation. Unlike contingency, constraint and template, all of which are deterministic, chance is distinguished by this stochasticity. Numerous statistical tools can help us generalize chance events, so they too are not beyond the bounds of scientific investigation.

Chaos, by contrast, often looks like chance, but is completely deterministic. Chaotic systems display an extreme sensitivity to their current state, and their longer-term dynamic behaviors often appear completely unexpected (Li & Yorke, 1975). Disease epidemics can exhibit chaotic dynamics, but their deterministic nature allows the development of analytical tools that enable short-term prediction (e.g., Sugihara & May, 1990).

Capriciousness, too, is deterministic, but here the order of events is critical. Capricious systems do not obey the "law of larger numbers"; things do not even out in the end. For instance, shuffling the order of variable selection schemes in simple population-genetic models leads to different outcomes (Lewontin, 1966), so it makes no sense to ask about average selection pressures or, indeed, even equilibria.

I call two further, related flavors of history *approach* and *turnover*. Both arise from a realization that the biological system may not be at equilibrium. Some systems do indeed have an equilibrium, but for various reasons take a long time to reach that point, or the equilibrium is easily disturbed and the system is not at the point much of the time. Whatever the case, the system spends most of its history approaching the equilibrium. The well-known mutation-selection balance of population genetics is an equilibrium at which the increase in frequency of a deleterious allele by mutation is balanced by the rate of removal by natural selection. For example, for a completely recessive allele arising at a rate μ in a population where homozygotes for this allele have their fitness reduced by an amount s, its equilibrium frequency is given by $\sqrt{\mu/s}$ (e.g., Hartl & Clark, 2007). But the glacial approach to this value means that observed frequency of the mutation in a real population may be very different. Indeed, the equilibrium may never be reached. Consequently, if one assumes that the frequency is in mutation-selection balance, the derivation of parameter estimates using this formula will be wrong.

Turnover is exemplified by the theory of island biogeography (MacArthur & Wilson, 1967), which regards the biodiversity of an area to be in a dynamic steady state, with the overall level roughly in balance while different species come and go. Here, the number of species may be at some sort of equilibrium (or close to one), determined by such things as the island's size and distance from a source population, but the species composition of the island keeps turning over. A similar steady state occurs in the neutral theory of molecular evolution, in which the various neutral alleles are effectively equivalent, all destined for eventual extinction even though the overall level of allelic variation is steady over the long term.

I have alluded to history as *construction* above. Construction focuses on the way in which properties of the system emerge as it is constructed or assembled over time. Moreover, this assembly may continue indefinitely, without any final equilibrium being reached. As an example, population-genetic simulations that simultaneously allow novel mutation and selection to act on populations over many generations (e.g., Spencer & Marks, 1988; Marks & Spencer, 1991) show that selectively maintained polymorphisms are surprisingly easy to construct, even though the parameters affording such stable polymorphisms are exceedingly rare in parameter space (Lewontin et al., 1978). Lewontin (1974) had dubbed an apparent mismatch between the theory and observations of widespread polymorphism "the paradox of variation". The incorporation of history into the theory goes some way to resolving Lewontin's paradox (Spencer & Marks, 1993).

Similarly, theoretical ecologist Peter Taylor showed that Lotka-Volterra models of multi-species ecosystems constructed by having species immigrate into existing communities one at a time resulted in far more species being present than an ahistorical analysis focused solely on equilibria (Taylor, 1988). These constructed communities exemplify the use of an historical approach to provide an explanation for a theoretical conundrum, namely the lack of a correlation between stability and complexity in model ecosystems first identified by physicist-turned-biologist Bob May (May, 1973).

History as construction is probably the flavor that has had the most impact on research in ecology and evolution. Certainly, a number of modelers in these disciplines have taken an overtly constructionist approach to their research questions. As a consequence, I spend more time discussing this flavor (in Chapters 8 and 9) and, reflecting the greater impact, the discussion is considerably more detailed about research results. Although it may feel like a "step-up" to the reader, my goal is to show how a well-developed research program that explicitly acknowledges history can make real strides, even if this progress is not straightforward. In some way, the use of history as construction could serve as a model for the practice of other flavors of history that have not been taken up to the same extent.

1.5 Concluding Remarks

To summarize, I contend that both the use of an historical approach and the recognition of the distinct historical flavors identified above provide real insight. The historical approach suggests testable hypotheses for otherwise hard-to-explain (or even previously unnoticed) observations. It reminds us, for example, to check whether or not the study system really is at equilibrium, which matters (as the mutation-selection balance example warns us) because the uncritical assumption that it is can lead to erroneous conclusions.

Moreover, distinguishing between the flavors (summarized in Table 1.1) is important because they imply different questions and assumptions. They focus on different aspects of the system (although, clearly, not all flavors will be present, and some are almost certainly more important than others). For instance, the foraging fish example assumes that the distribution of consumers is at equilibrium, but the models of constructed polymorphisms do not make the same presumption about the constituent allele frequencies. Capriciousness reminds us that the order of events may not just even out their effects, whereas chaos implies that the unexpected need not be due to chance. This book argues that an explicit consideration of various historical flavors will often lead to a deeper, more nuanced understanding of ecological and evolutionary biology. History matters, but in different ways.

Table 1.1 The Flavors of History

Flavor	Brief Definition, *Exemplar(s)*	Chapter
Contingency	Occurrence of events is conditional on the properties of the system *Competitive feeding fish*	3
Constraint	Some possible events are ruled out by previous events *Genome reduction and the "minimal gene set"*	4
Template	Particular events occur on an historical background (e.g., phylogeny) *Coevolution of wing lice and pigeons*	5
Chance	Events occur randomly; the only truly stochastic flavor *Genetic drift*	6
Chaos	Events occur with no discernable pattern that is not random *Infectious disease epidemics*	6
Capriciousness	Reversing (or shuffling) the order of events does not even out *Reversing variable selection pressures*	6
Approach	The system gets near, but seldom/never arrives at equilibrium *Mutation-selection balance*	7
Turnover	Elements of the system continually change *Island biogeography; neutral theory of molecular evolution*	7
Construction	The system assembles over time *Polymorphism construction; community assembly*	8, 9

References

Beatty J. 1987. Weighing the risks: Stalemate in the classical/balance controversy. *Journal of the History of Biology* 20:289–319.

Cain J. 1994. Ernst Mayr as community architect: Launching the Society for the Study of Evolution and the journal *Evolution*. *Biology and Philosophy* 9:387–427.

Conference on Concepts of Biology. 1958. Concepts of Biology. Condensed Transcript of the Conference. *Behavioral Science* 3:103–195.

Darwin C. 1859. *On the Origin of Species by Means of Natural Selection.* London: John Murray.

Estes J. A., Vermeij G. J. 2022. History's legacy: Why future progress in ecology demands a view of the past. *Ecology* 103:e3788.

Fretwell S. D., Lucas H. L. 1970. On territorial behavior and other factors influencing habitat distribution in birds. I. Theoretical development. *Acta Biotheoretica*, 19:16–36.

Hartl D. L., Clark A. G. 2007. *Principles of Population Genetics.* 4th ed. Sunderland, Mass.: Sinauer.

Houston A. I., McNamara J. M. 1988. The ideal free distribution when competitive abilities differ: An approach based on statistical mechanics. *Animal Behaviour* 36:166–174.

Johnson K. P., Clayton D. H. 2003. Coevolutionary history of ecological replicates: Comparing phylogenies of wing and body lice to columbiform hosts. Pages 262–286 in *Tangled Trees:*

Phylogeny, Cospeciation, and Coevolution, edited by R. D. M. Page. Chicago: University of Chicago Press.

Kennedy M., Gray R. D. 1993. Can ecological theory predict the distribution of foraging animals? A critical analysis of experiments on the ideal free distribution. *Oikos* 68:158–166.

Kingsland S. E. 1985. *Modeling Nature: Episodes in the History of Population Ecology*. Chicago: University of Chicago Press.

Lewontin R. C. 1966. Is nature probable or capricious? *Bioscience* 16:25–27.

Lewontin R. C. 1974. *The Genetic Basis of Evolutionary Change*. New York: Columbia University Press.

Lewontin R. C., Ginzburg L. R., Tuljapurkur S. D. 1978. Heterosis as an explanation for large amounts of polymorphism. *Genetics* 88:149–170.

Li T.-Y., Yorke J. A. 1975. Period three implies chaos. *American Mathematical Monthly* 82:985–992.

MacArthur R. H., Wilson E. O. 1967. *The Theory of Island Biogeography*. Princeton: Princeton University Press.

Marks R. W., Spencer H. G. 1991. The maintenance of single-locus polymorphism. II. The evolution of fitnesses and allele frequencies. *American Naturalist* 138:1354–1371.

May R. M. 1973. *Stability and Complexity in Model Ecosystems*. Princeton: Princeton University Press.

Orwell G. 1945. *Animal Farm: A Fairy Story*. London: Secker & Warburg.

Parker G. S., Sutherland W. J. 1986. Ideal free distributions when individuals differ in competitive ability: Phenotype-limited ideal free models. *Animal Behaviour* 34:1222–1242.

Ruse M. 1996. *Monad to Man: The Concept of Progress in Evolutionary Biology*. Cambridge, Mass.: Harvard University Press.

Servedio M. R., Brandvain Y., Dhole S., Fitzpatrick C. L., Goldberg E. E., Stern C. A., Van Cleve J., Yeh D. J. 2014. Not just a theory—The utility of mathematical models in evolutionary biology. *PLoS Biology* 12:e1002017.

Spencer H. G. 1998. Onward and upward. Symposium review of *Monad to Man: The Concept of Progress in Evolutionary Biology*, by Michael Ruse, Harvard University Press. *Metascience* 7:56–61.

Spencer H. G. 2020. Beyond equilibria: The neglected role of history in ecology and evolution. *Quarterly Review of Biology* 95:311–321.

Spencer H. G., Kennedy M., Gray R. D. 1995. Patch choice with competitive asymmetries and perceptual limits: The importance of history. *Animal Behaviour* 50:497–508.

Spencer H. G., Marks R. W. 1988. The maintenance of single-locus polymorphism. I. Numerical studies of a viability selection model. *Genetics* 120:605–613.

Spencer H. G., Marks R. W. 1993. The evolutionary construction of molecular polymorphisms. *New Zealand Journal of Botany* 31:249–256.

Sugihara G., May R. 1990. Nonlinear forecasting as a way of distinguishing chaos from measurement error in time series. *Nature* 344:734–741.

Taylor P. J. 1988. The construction and turnover of complex community models having generalized Lotka-Volterra dynamics. *Journal of Theoretical Biology* 135:569–588.

2

The Equilibrium

2.1 Equilibria in Ecology and Evolution

Biological equilibria are ubiquitous in ecological and evolutionary research (and, indeed, in much of biological research more broadly). Often their presence is hidden in plain sight: the discussion of the scientific issues makes an assumption about an equilibrium that is seldom explicit. The competitive foraging fish example from Chapter 1, for instance, is discussed in terms of the distribution of competitors, but usually there is no mention that this distribution is an equilibrium. More generally, analyses are carried out with the implicit presumption that the system is at equilibrium: the numerous calculations of the frequency of a deleterious mutation in a population are a case in point. But what do we really mean when we talk about an equilibrium?

Box 2.1 Terminology

Some key words in mathematical models in ecology and evolution, with examples drawn from the standard population-genetic model of heterozygote advantage (see Box 2.2).

Parameter	A quantity in a model that remains constant (e.g., the selection coefficients s and t)
Parameter space	The range of possible values the parameters may take (e.g., the interval (0, 1] representing the possible values of t; or the square given by $\{(s, t): 0, < s, t \leq 1\}$)
Variable	A quantity in a model that changes over time (e.g., the allele frequency p)
State space	The range of possible values the variables may take (e.g., the unit interval [0, 1] representing possible values of p). A description of the possible configurations of the system. Sometimes called phase space, especially when several variables are present. The way a two- or three-variable system changes over time is often portrayed by a trajectory plotted in two- or three-dimensional state space.

Beyond Equilibria. Hamish G. Spencer, Oxford University Press. © Hamish G. Spencer (2025).
DOI: 10.1093/oso/9780192858993.003.0002

The models examined in this book are usually *dynamical* models: they describe the behavior of some biological system as it changes over time. A model of a dynamical biological system is considered to be at equilibrium when the variables in the system (the things that potentially change) do not change over time. Perhaps the simplest sort of equilibrium is when nothing at all is changing: allele frequencies (the common variables in population-genetic models) are at values determined by heterozygote advantage in a constant environment, for instance (see Fig. 2.1). Or, as an ecological example, under the simple logistic model of population size there is an equilibrium at which the population size is at its so-called carrying capacity, K. Of course, we can quibble about the small variations in allele frequency due to random genetic drift, or the possibility that the population size may fluctuate around K, but these considerations are secondary to what we have decided is important about the model and, hence, how it is analyzed. The system is currently at rest, unchanging over time.

Our simple definition can, however, obscure a host of interesting phenomena. How would we describe a system in which the variable of interest is constant, at least in the long term, and yet underlying change is ongoing? The number of species on an island in a model of island biogeography, or the level of heterozygosity in a constant-sized population under the neutral hypothesis come to mind (see Fig. 2.2). The actual species and alleles in the respective populations are being replaced over time, and yet their numbers, as well as derived measures, such as heterozygosity, reach a steady state, a value that is constant over long periods of time, which we surely would describe as an equilibrium.

If it is important to distinguish between these two sorts of equilibria, we can describe the first (i.e., that in the models of heterozygote advantage or logistic population growth) as a *static* equilibrium. The second (i.e., that in the models of island biogeography or neutral variation) is a *dynamic* equilibrium, or, if we want to emphasize the underlying active change in the system, we can label it a *steady state*.

There are further kinds of possible outcomes that are not truly equilibria, but which nevertheless have a certain regularity. In two dimensions, for example, a *limit cycle* occurs when our variable of interest continues to change, but only takes on certain values in a predictably repeatable way (see, e.g., Fig. 2.3). The variable cycles around the set of possible values, a set that is possibly infinite. And it is even possible for a system not to have any equilibria. For example, some systems display chaotic behavior, in which the variable keeps changing to new values that seem to show no regard for their previous values, although it is, in fact, completely predictable (see section 6.3).

16 • *Beyond Equilibria*

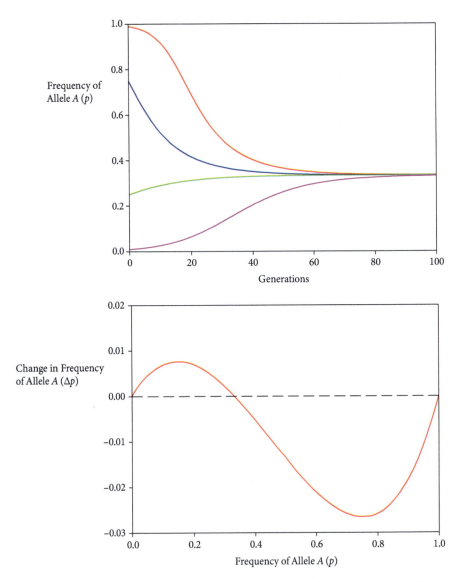

Fig. 2.1 Equilibrium in a model of heterozygote advantage. *AA* homozygotes have a fitness of 0.8 compared to *Aa* heterozygotes; *aa* homozygotes have a relative fitness of 0.9. Top: Frequency of allele *A* over time (p) for four different initial values. Given the fitnesses in the underlying model, the system comes to rest at a value of $p = ⅓$ irrespective of its starting point. Bottom: Change in allele *A*'s frequency (Δp) as a function of p. The equilibrium corresponds to $\Delta p = 0$ (the horizontal dashed line). Note that there are, in fact, three equilibria: $p = 0$ and $p = 1$ are unstable; only $p = ⅓$ is stable.

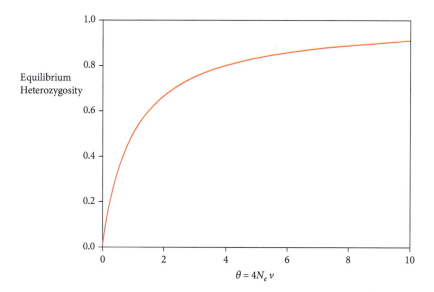

Fig. 2.2 Expected heterozygosity at the steady state predicted by the neutral model of molecular evolution. N_e is the effective population size; v is the neutral mutation rate; the compound parameter $\theta = 4N_e v$, which increases linearly with both population size and mutation rate.

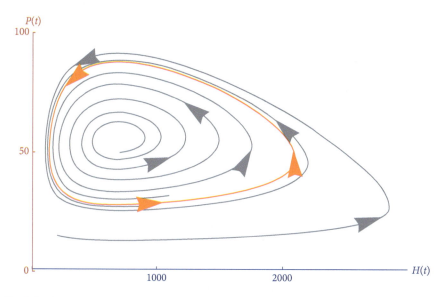

Fig. 2.3 The trajectories of a predator-prey model exhibiting a stable limit cycle, plotted in state space. The x-axis shows the numbers of prey; the y-axis the numbers of predators. A system starting with 700 prey and 50 predators iterates with gradually more extreme population sizes, closer and closer to the orange line (the limit cycle). Systems initiated with values outside the limit cycle (e.g., 200 and 15, respectively) also approach it, but with gradually less extreme sizes.

Box 2.2 Finding the Equilibrium of a Model

In principle, finding the value of the independent variable at an equilibrium in a model is simple: you take the equation(s) describing the change in the variable(s) of interest, equate it (them) to zero (since an equilibrium corresponds to no change) and solve for the variable(s).

Consider, for instance, the standard population-genetic model of heterozygote advantage. The difference equation describing the change in allele frequency from one generation to the next in an infinite population with two alleles, A and a, is given by

$$\Delta p = \frac{p(1-p)(t-p(s+t))}{1-sp^2-t(1-p)^2},$$

where p is the frequency of the A allele and s and t are the selection coefficients against the AA and aa genotypes. Setting $\Delta p = 0$ and solving for p gives three solutions: $p = 0$, $p = 1$ and $p = t/(s+t)$. The first two solutions, known mathematically as trivial solutions, correspond to the extinction and fixation of A; when there is no genetic variation, selection changes nothing. The third solution is the one of most theoretical interest, since both alleles are present in the population; the population is polymorphic.

As a second example, we can examine the well-known logistic model for population growth, described by the differential equation

$$\frac{dN}{dt} = rN\left(1 - \frac{N}{K}\right),$$

where N is the population size, t is time, $r > 0$ is the growth rate and K is the carrying capacity. There are clearly two solutions to $dN/dt = 0$, one trivial one when $N = 0$ and the population is extinct, and one when $N = K$ and the population is at the carrying capacity.

For more complicated models, however, especially ones with several variables, the solution may be far more difficult or even analytically impossible (albeit sometimes with an approximate or numerical solution). And, finally, there may not be a solution at all; the system may never stop changing, the variables never ever settling down. Such an outcome is often seen in systems with history as chaos or chance.

2.2 Stability of Equilibria

One of the reasons we are interested in biological equilibria is because we would like to know about the long-term behavior of our dynamical system. In the examples above, the variable of interest moves toward an equilibrium. The allele frequency under heterozygote advantage iterates toward $p = t/(s+t)$; the logistic population converges to the carrying capacity, K; heterozygosity under neutrality stabilizes at

$\theta/(1 + \theta)$. But even in these simple cases, there are other equilibria (admittedly trivial and rather uninteresting ones) to which the system does not converge. For instance, in the case of heterozygote advantage, both the extinction of allele A and its fixation— $p = 0$ and 1—are also equilibria. How can we characterize the equilibria in our model in a way that captures these different outcomes? The answer is that we use the concept of stability (see Fig. 2.4).

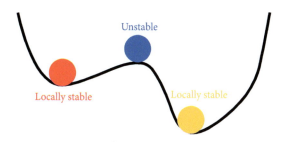

Fig. 2.4 Different classes of stability displayed by a ball resting on the black surface. Small perturbations from either of the two locally stable equilibria allow the ball to return to that equilibrium, whereas any change in position from the neutrally stable equilibrium causes the ball to move further away (to either of the stable equilibria). Note that a larger move away from a locally stable equilibrium may mean the ball does not return, so neither is globally stable. The red locally stable equilibrium, however, is metastable, since a large push to the right will cause the ball to drop to the lower ("better") yellow equilibrium. A neutrally stable equilibrium would correspond to a flat interval on the black surface.

An equilibrium is said to be *stable* if, for values of the variable close to the equilibrium value, the system converges to that equilibrium. Sometimes (e.g., Lewontin, 1969; May, 1974) this property is called *local* or *neighborhood stability*, emphasizing that this property pertains to variable values close to the equilibrium. An instructive way to think of this property is to ask what would happen if the system at equilibrium were disturbed in a small way: would it return to the same equilibrium? In the case of heterozygote advantage, the polymorphic equilibrium is stable. We can see in the top panel of Fig. 2.1 that, for all initial non-trivial frequencies of A (i.e., $0 < p < 1$), the value of p converges over time to that equilibrium. The bottom panel implies the same conclusion, but in a different fashion. It shows that for frequencies of A below the equilibrium, the change in frequency is positive; A becomes commoner, moving closer to the polymorphic equilibrium. In symbols, for all $p < t/(s + t)$, $\Delta p > 0$. Similarly, if $p > t/(s + t)$, $\Delta p < 0$, and the system again iterates closer to the equilibrium, with the frequency of A decreasing. See Box 2.3 for a more formal argument.

In fact, the polymorphic equilibrium under heterozygote advantage is *globally stable*, since the frequency of A converges to that equilibrium from all biologically relevant initial values (i.e., $0 < p < 1$). In some more complicated models, however, convergence to a stable equilibrium is limited to a subset of possible initial values; those *locally stable* equilibria are not globally stable. Some locally stable equilibria may also be described as *metastable*, which applies when the system variable is altered away from that equilibrium and the system converges to some other equilibrium that is in some way "better" (e.g., an equilibrium in a population-genetic model with higher mean fitness).

Conversely, if the system moves away from an equilibrium no matter how close you are, it is said to be *unstable*. For example, the extinction and fixation equilibria in the model of heterozygote advantage are unstable, since for any allele frequency close to either of them (but not actually zero or one), the system iterates away from them, toward the globally stable polymorphic equilibrium. Similarly, the trivial equilibrium in the logistic model is unstable since, for even the smallest (but non-zero) values of N, population growth is positive and hence N becomes larger, diverging from 0. More formally, for N close to 0, $dN/dt \approx rN$, which is positive since both r and N are positive.

Box 2.3 Classifying the Equilibrium of a Model

Deciding if the equilibrium of a model is locally stable or unstable involves determining the dynamical behavior for values of the variable(s) close to the equilibrium. We want to know if a small displacement of the variables from their equilibrial values dies away over time (meaning local stability) or whether this perturbation grows (and hence the equilibrium is unstable).

In some cases, we can answer this question directly. Continuing our heterozygote advantage example from Box 2.2, for example, we can analyze the expression for change in allele frequency,

$$\Delta p = \frac{p(1-p)(t - p(s+t))}{1 - sp^2 - t(1-p)^2},$$

with little trouble. The denominator is always positive (for the range of biologically relevant parameter values, $0 < s, t \leq 1$, and variable values $0 < p < 1$). So, for example, the trivial equilibrium $p = 0$ can be classified by noting that the three terms in the numerator are, for values of p just larger than 0, all positive (very small but positive, ~1 and ~t, respectively). Hence, for small values of p close to 0, Δp is also positive: p increases, moving away from the trivial equilibrium, which is thus unstable.

Similarly, for the equilibrium $p = t/(s+t)$, stability is determined by the sign of the third term of the numerator. For values of p just less than the equilibrium, it is positive; for values just greater, it is negative. Hence, any displacement dies away, the value of p moving closer back toward the locally stable equilibrium.

Alternatively, we can use some elementary calculus, evaluating the derivative (with respect to the variable) of the above expression evaluated at the equilibrium. When this value is negative, the equilibrium is locally stable; when it is positive, it is unstable. At

Box 2.3 *Continued*

$p = 0$, this approach gives $t/(1 - t)$, which is clearly positive, and the equilibrium unstable; at the polymorphic equilibrium, the expression is more complicated algebraically, but it can be seen that it is always negative and the equilibrium thus locally stable.

Our logistic example from Box 2.2 gives a derivative of $r - 2rN/K$, which when $N = 0$ is $r > 0$. Hence, the trivial equilibrium is unstable. When $N = K$, the derivative is $-r < 0$; the carrying capacity is locally stable. This use of calculus will fail to resolve the matter, however, if the derivative is zero. In such cases, there are other techniques (see Roughgarden, 1979; Edelstein-Keshet, 1988; Otto & Day, 2007).

For systems with more than one variable, the second approach involving calculus is often easier. The equations describing the behavior of each variable in terms of all the others can be viewed as a single vector-valued function, and the dynamics near any equilibrium are determined by the leading eigenvalue of the *Jacobian*, the (square) matrix of the first-order partial derivatives, evaluated at that equilibrium. This topic is described in detail for models in discrete and continuous time in Roughgarden (1979), Edelstein-Keshet (1988) and Otto and Day (2007).

Global stability is a far more demanding criterion, and there is often no simple analytical method for its determination. Ad hoc approaches (such as examining Fig. 2.1 for heterozygote advantage) can sometimes work, and for "well-behaved" systems (which usually means that the equations and their derivatives are continuous) there may be applicable theorems that show global stability. In general, however, for models with any degree of complexity, global stability is unlikely.

In between stable and unstable equilibria are *neutrally stable* equilibria. In this class, for values of the variable close to the equilibrium, the system simply remains at those values. In effect, neutrally stable equilibria act like a region of stable equilibria, but they are rare in ecological and evolutionary models and their practical importance is rather limited.

Distinguishing between unstable, metastable, neutrally stable and globally and locally stable equilibria in a model can be difficult in practice. Nevertheless, there are a number of analytical techniques for doing so (see Box 2.3). The interested reader is encouraged to consult a more specialist text for a more detailed discussion of this topic (e.g., Roughgarden, 1979; Edelstein-Keshet, 1988; Otto & Day, 2007).

Similarly, there are more sophisticated mathematical methods for working out how quickly one approaches an equilibrium, given that it is stable. For instance, the logistic model of population growth adumbrated in Box 2.2 has a globally stable equilibrium at $N = K$, but it is not immediately obvious, perhaps, how rapidly that equilibrium might be reached. Nevertheless, larger values of r, the growth rate, surely mean that the population grows faster and, indeed, that is the case. (Very large values of r can mean N overshoots the equilibrium, however, which may lead to subsequent large declines in the population size or even extinction, but for most biologically relevant growth rates and population sizes, the conclusion holds.) Again, the interested reader may find advanced treatments useful.

2.3 Structural Stability

Classifying the stability of equilibria involves examining the behavior of the model when the system is close to an equilibrium, perhaps as a result of some perturbation to the values of the variables describing the state of the system. A rather different concept of stability concerns the structure of the model. For example, we might want to know how stable it is to variations in its parameters (as opposed to variables). Perhaps, even more fundamentally, we could ask how stable our model is to differences in its structure.

As ecologists and evolutionists, we are interested in such questions because we know our models are but caricatures of the biological system we are studying. Yes, of course, some of us are more interested in the esoteric mathematical properties of the systems of equations we are using in our modeling, but many of us are motivated by how our modeling might pertain to the real world. Consequently, it makes sense to ask how robust our conclusions might be to the uncertainty in parameter estimates or differences in the form of the equations we are using to describe the biological phenomena.

These questions are a little vague, though; it is not exactly clear what we mean by stability. Even in the very simplest models, the value of any equilibrium is likely to change if the model's parameters change. Hence, in defining what we mean by a structurally stable model, we cannot demand that the model has the exact same equilibria. With heterozygote advantage, for instance, the equilibrium value for the allele frequency of $p = t/(s + t)$ is altered by any change in the value we assign to the parameter t. The logistic model, however, has its globally stable equilibrium of $N = K$ unaffected by the value of the growth rate, r. Nevertheless, it behaves differently if r changes; the equilibrium may be the same, but the rate of approach is not.

Without going into too many mathematical details, the idea of structural stability is that small deviations in parameters should result in only small deviations in equilibrium values of the variables. We will see that this is not always the case (see, e.g., Box 2.4), even in some rather simple models. In our heterozygote advantage example, though, we can see at an intuitive level that small changes in the value of t will usually lead to small changes in the value of the polymorphic equilibrium. And we might also expect that varying the parameters for a structurally stable model might preserve some of that model's properties, such as the number of equilibria.

Box 2.4 An Example of a Model with Cycling and an Illustration of Structural Stability

Gavrilets (1998) proposed a single-locus diallelic population-genetic model in which both the mother's and the individual's own genotype affected fitnesses. Suppose $i = 1, 2$ and 3 correspond to the genotypes AA, Aa and aa, respectively, and that w_{ij} is the fitness of individuals of genotype i with genotype j mothers. If x, y and z are the post-selection frequencies

Box 2.4 *Continued*

of adults with respective genotypes AA, Aa and aa, then the recursion equations for these frequencies in the following generation are

$$\bar{w}x' = w_{11}px + \frac{1}{2}w_{12}py$$

$$\bar{w}y' = w_{23}pz + \frac{1}{2}w_{22}y + w_{21}qx$$

$$\bar{w}z' = \frac{1}{2}w_{32}qy + w_{33}qz$$

in which $p = x + \frac{1}{2}y$, $q = 1 - p$ and $\bar{w}$ is the population's mean fitness, the sum of the right-hand sides of the three recursions (Spencer, 2003).

This model can exhibit cycling. When $w_{11} = 0.028$, $w_{12} = 0.966$, $w_{21} = 0.116$, $w_{22} = 0.011$, $w_{23} = 0.886$, $w_{32} = 0.614$ and $w_{33} = 0.042$, (x, y, z) alternates between $(0.2769, 0.6315, 0.0916)$ and $(0.5608, 0.1956, 0.3306)$, as shown in part a of the diagram, in which the solid line is x and the dotted line y.

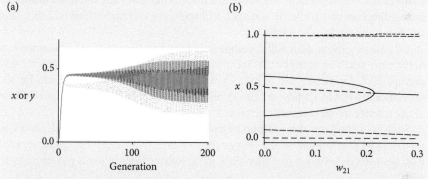

Part b, known as a bifurcation diagram, allows us to understand aspects of the structural stability of the model. The graph shows the equilibrium values of x as a function of w_{21}, with solid lines indicating stable equilibria and dotted lines unstable ones. For $w_{21} > \sim 0.214$, just a single equilibrium is stable; the remaining ones are all unstable. As w_{21} is gradually reduced from 0.3 to 0.214, the equilibrium value of x slowly increases; this consistency is structural stability. As we reduce w_{21} below 0.214, however, a stable two-cycle (as shown in a) bifurcates from this stable equilibrium, which becomes unstable; this change corresponds to structural instability. See Spencer (2003) for further details. See also section 6.3.

Changing the form of equations poses an even stronger challenge to any concept of stability. After all, even small changes in the equations are likely to lead to significant differences in the properties of the models; we cannot expect the values or even numbers of equilibria to remain the same. Nevertheless, some models can behave very similarly for certain ranges of the variables. For example, the equation for exponential growth of a population, $dN/dt = rN$, is very like the logistic for some values of N and the two models share the trivial equilibrium of $N = 0$ and dN/dt is very similar in both models when N is small.

2.4 Stochastic Versus Deterministic Models

The models discussed above are all *deterministic*: the behavior of the model is fully determined by the relevant equations. By contrast, *stochastic* models have an element of randomness. In population genetics, for example, models incorporating the effects of genetic drift are stochastic. In such models, the short-term behavior is not completely predictable, although there may well be a probability distribution of possible short-term changes in the variables and the longer-term consequences, possibly including equilibria, may well be fully predictable.

The addition of stochasticity to deterministic models illustrates the importance of understanding the equilibrium behavior of these original models. For example, a model for the action of heterozygote advantage in a finite population adds genetic drift to the equations of Box 2.2. The consequence of genetic drift is to randomly perturb the allele frequencies each generation, by a small amount in a random direction. If the system were at the polymorphic equilibrium, for example, drift is mirroring the thought experiment we carry out when deciding whether this equilibrium is locally stable or not. We know from the above discussions that after any small deviation from the equilibrium ($p = t/(s + t)$), selection will move the system back toward this value, since it is, in fact, locally stable.

Of course, genetic drift will continue to act, and so the extended model now can never truly reach an equilibrium: it is forever changing. Nevertheless, the long-term average allele frequency will remain close to the deterministic equilibrium (at least for values of s not too different from t), and so the simpler model still captures much of the essence of the stochastic version.

Such robustness to the addition of randomness is not always the rule, however: stochasticity can fundamentally change the outcome of a model. For example, selection against a deleterious recessive allele in an infinite population is governed by the equation

$$q' = \frac{q(1 - sq)}{1 - sq^2}, \tag{2.1}$$

where q is the frequency of the deleterious allele, q' that in the following generation and s is the selection coefficient against the recessive homozygote. The frequency of this allele declines each generation, as shown in Fig. 2.5. But this decrease becomes smaller over time: when the allele is rare, the vast majority of those alleles will be found in heterozygotes, sheltered from selection. Only increasingly rare homozygotes are targeted and the efficacy of selection in improving the mean fitness of the population declines. The long tail in Fig. 2.5, which never reaches the equilibrium of zero (i.e., extinction of the allele), shows this weakening of the selective process.

The addition of genetic drift to this model (i.e., assuming a finite population) has a major consequence. Genetic drift results in small, random changes to the allele frequency and, when the allele is rare, can result in its extinction. Thus, in comparison to the deterministic model, which never quite reaches equilibrium, the stochastic model will do so, but not in a predictable number of generations.

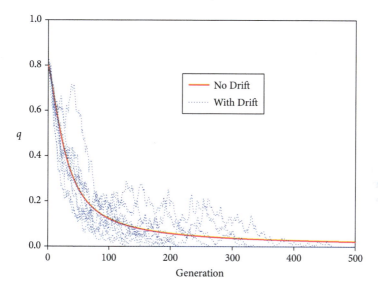

Fig. 2.5 Allele frequency (q) of a deleterious recessive with selection coefficient $s = 0.1$, starting at $q = 0.8$. The equilibrium value of q is zero. The red line shows the change under the deterministic model of Equation 2.1; the dotted blue lines show the results of 10 stochastic simulations with genetic drift, coming from the assumption of a finite population of size 500.

2.5 Multiple Equilibria

In much of the above discussion, we have assumed (sometimes without even being explicit) that the system we are examining has just one equilibrium ("the" equilibrium). Nevertheless, recall that a stable equilibrium may only be locally stable, which implies that for values of the model's variable(s) sufficiently far away from this equilibrium, the system will not return there. Consequently, it must converge to another locally stable equilibrium or not converge at all.

Indeed, for many ecological and evolutionary models, distinct equilibria are possible, perhaps depending on the initial values of the variables or possibly depending on random events as the system develops. In population genetics, the standard model for heterozygote disadvantage has the same equilibria as that of heterozygote advantage (see Fig. 2.6), but the stability of these equilibria is reversed. Hence a population starting off at a lower allele frequency will lose that allele; a population initially at a higher frequency will fix that allele. The system of competitive foraging fish from Chapter 1 (see section 1.2) is an example of how different equilibria—the four different ratios—may be reached, depending on how the system builds up over time.

Lewontin (1969) was one of the first to raise this issue in ecology, arguing that if multiple equilibria were possible, then an adequate explanation of why a particular equilibrium was achieved (rather than some alternative) necessitated taking history into account. The long-standing theories about ecological succession terminating in a unique climax community clearly envisaged a single globally stable equilibrium. The following decade witnessed a plethora of experimental investigations into whether or not such a picture was accurate.

2.6 The Importance of Different Outcomes

The ideas above are illustrated rather well in a recent investigation of models of cheats and cooperators. Populations of cooperative organisms are vulnerable to invasion by cheats, who benefit from the cooperative actions of the resident population, but do not contribute their fair share. Liu et al. (2023) noted that the dynamics of the proportion of cheats in a population in both experimental and theoretical studies were extremely varied. In some cases, cheats died out; in others they were fixed, leading to the elimination of cooperation. These two cases correspond to what are known as the trivial equilibria of extinction and fixation. In yet other cases, the system soon reached a stable equilibrium at which the proportions of both cheaters and cooperators remained constant, and a fourth possibility was a cyclical oscillation, with the frequencies of cheating forever fluctuating. The obvious questions are about whether or not the same biological mechanisms underlaid this diversity and what shaped these differences.

Liu et al. (2023) constructed a model incorporating four different life-history features: periodic reductions in population sizes, frequency- and density-dependent cheater growth rates and population structure. In many circumstances, their model converged to an equilibrium (sometimes one of the trivial ones), but a combination of periodic population bottlenecks and cheater growth rates being dependent on the density of cooperators could generate oscillations, the size of which could be affected by frequency dependence. Stochastic effects arising from incomplete mixing among subpopulations in a structured meta-population interacted further to produce longer-term oscillations (i.e., those longer than the time between bottlenecks). The richness of the possible dynamics is evident in part because of an awareness of different outcomes, equilibrial and otherwise. And, of course, these differences matter practically. As Liu et al. (2023) pointed out, control of parasites, viruses and other cheaters will depend on which life-history factors can be manipulated by a proposed treatment or other intervention.

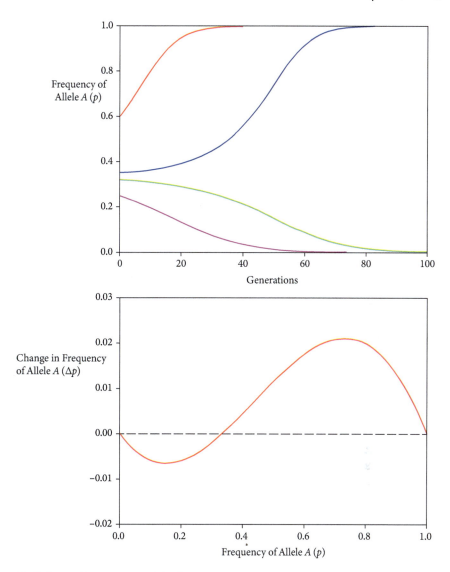

Fig. 2.6 Equilibrium in a model of heterozygote disadvantage. *AA* homozygotes have a fitness of 1.2 compared to *Aa* heterozygotes; *aa* homozygotes have a relative fitness of 1.1. Top: Frequency of allele *A* over time (*p*) for four different initial values. Given the fitnesses in the underlying model, the system comes to rest at a value of $p = 0$ or 1, depending on its starting point. Bottom: Change in allele *A*'s frequency (Δp) as a function of *p*. The equilibrium corresponds to $\Delta p = 0$ (the horizontal dashed line). Note that there are, in fact, three equilibria: $p = 0$ and $p = 1$ are locally stable; whereas $p = \frac{1}{3}$ is unstable.

2.7 Concluding Remarks

Understanding the behavior of a mathematical model (or even a verbal model) often begins with an analysis to find the equilibrium of that model. It is possible, though, that there are no equilibria or multiple equilibria. Moreover, each equilibrium can be locally or globally stable, unstable or metastable. Mathematical techniques often provide the tools for such analyses. Nevertheless, as the rest of this book argues, we need to look beyond the equilibrium if we want to fully understand our model and the biology that inspired it.

References

Edelstein-Keshet L. 1988. *Mathematical Models in Biology*. Boston, Mass.: McGraw-Hill.

Gavrilets S. 1998. One-locus two-allele models with maternal (parental) selection. *Genetics* 149:1147–1152.

Lewontin R. C. 1969. The meaning of stability. Pages 13–24 in *Diversity and Stability in Ecological Systems*. Upton, NY: Brookhaven National Laboratory.

Liu M., Wild G., West S. A. 2023. Equilibria and oscillations in cheat–cooperator dynamics. *Evolution Letters* 7:339–350.

May R. M. 1974. *Stability and Complexity in Model Ecosystems*. Princeton, NJ: Princeton University Press.

Otto S. P., Day T. 2007. *A Biologist's Guide to Mathematical Modeling in Ecology and Evolution*. Princeton, NJ: Princeton University Press.

Roughgarden J. 1979. *Theory of Population Genetics and Evolutionary Ecology: An Introduction*. New York, NY: Macmillan.

Spencer H. G. 2003. Further properties of Gavrilets' one-locus two-allele model of maternal selection. *Genetics* 164:1689–1692.

3

Contingency

3.1 The Flavor of Contingency

Contingent events depend on what has previously occurred and how those past events have impacted the system's properties. Surely, in ecological and evolutionary research, contingency is the most frequent of history's flavors. What occurs in a biological system depends on a slew of past events, so many and so wide ranging that it may appear that any form of generalization is impossible. In this chapter I argue that this despairing view of contingency is unwarranted. Contingent events in ecology and evolution can be studied systematically, often using mathematical tools. They can be modeled, generalizations can be made and, sometimes, even the future can be predicted. I illustrate my claims with examples drawn from introduction ecology, foraging theory and population genetics.

Moreover, in my view, contingency provides us with information about a system, often suggesting the direction for research and hypotheses to test. Such a process even occurs in our everyday lives. For instance, people's decisions are normally contingent on what has already transpired, happenings that add to the evidence base informing the decision maker. But let me return to ecology and evolution.

Undoubtedly, the commonest depiction of the role of history in ecology and evolution is *contingency*: the occurrence of events is conditional on the properties of the system in question. In the historical studies of the foraging fish (section 1.2), for instance, each individual entering the system chose the marginally more profitable site, this decision being contingent on the previous choices of the fish that had already arrived. The ahistorical bias of Kingsland's ecological theorists (section 1.3) assumes that past contingency renders any generalizing impossible. But, as we have seen a mathematical model with contingency does allow generalizations, inferences that assist us to explain real-world observations.

Contingent events depend critically on what has already happened. This dependency arises largely because those past events have affected the system's properties in a way that alters the suite of potential responses to the current inputs and the respective likelihoods of these possible responses. Every biological (and, indeed, non-biological) system has some degree of contingency. But if the level of contingency were too great,

Beyond Equilibria. Hamish G. Spencer, Oxford University Press. © Hamish G. Spencer (2025).
DOI: 10.1093/oso/9780192858993.003.0003

very little scientific investigation could proceed. We would be stuck with a catalog of special cases, little generalization and no predictive ability.

One of the reasons mathematical modeling is so useful in dealing with contingency is that we can re-run the experiment, avoiding the anecdotal nature of unique, contingent events. This repetition may be in the form of simulations (as in the foraging fish) or in the use of analytical techniques (such as those used to find and characterize equilibria). Repetition may allow us to average-out the contingent effects that impede useful generalizations. In some circumstances, empirical investigations can make use of naturally replicated real-world events. Koch et al. (2022), for example, used the repeated evolution of two distinct ecotypes in the marine snail *Littorina saxatalis* on three Swedish islands to show which parts of the genome consistently underlaid the phenotypic variation and which parts had responded differently among islands, in a more contingent manner.

I start this chapter by discussing a non-model example of contingency and how its consequences might be investigated. In short, we employ various standard scientific procedures to reduce the degree of contingency, leaving us with a system we can explain (or at least explore further). In some ways, however, the resolution of this problem is a distraction from my main message, which is how to incorporate contingency into our understanding in a systematic way.

3.2 Passerine Introductions in New Zealand

Ornithologists recognize New Zealand's living avifauna as globally special. There are six endemic families and more species of breeding seabirds than anywhere else (del Hoyo, 2020). One less well-known aspect, though, is the large number of passerine introductions, many of which have been successful in establishing breeding populations (Thomson, 1922; Gill et al., 2010; Pipek et al., 2015). For example, four species of buntings (Fig. 3.1) have been imported from Europe and released, each with a different outcome. The yellowhammer (*Emberiza citrinella*) is now common and widespread in agricultural areas throughout the country. The cirl bunting (*E. cirlus*), also occurring on farmland, is much rarer, commoner in the north of the South Island, but with only scattered populations elsewhere. The ortolan bunting (*E. hortulana*) successfully bred in the southern North Island, but subsequently died out. Finally, the common reed bunting (*E. schoeniclus*) disappeared soon after the release of fewer than a dozen birds in the southern South Island.

What are we to make of this varied history? Four closely related species, yet four different outcomes; four special cases, each one contingent on a myriad of particularities. Surely any model of population sizes for these introduced birds would contain so many variables and parameters difficult to measure that any generalizing would be precluded. Moulton et al. (2012) seem to epitomize this view, criticizing analyses that concluded that the numbers of birds released was critical and writing,

It is likely that the outcome of passerine bird introductions to New Zealand depended on species characteristics, site characteristics, and human decisions more than on a simple summing of the numbers introduced.

Fig. 3.1 From left to right: Yellowhammer (*Emberiza citrinella*); cirl bunting (*E. cirlus*); ortolan bunting (*E. hortulana*); common reed bunting (*E. schoeniclus*).

Photo of yellowhammer by Andy Morffew. Photo of cirl bunting by Paco Gómez. Both cropped and resized under Creative Commons Attribution 2.0 Generic license. Photo of ortolan bunting by Pierre Dalous. Cropped and resized under Creative Commons Attribution-Share Alike 3.0 Unported license. Photo of common reed bunting by Smudge 9000. Cropped and resized under Creative Commons Attribution-Share Alike 2.0 Generic license.

But surely, we would want to do better. For instance, could we have predicted which species would have become established? More specifically, if an attempt were made to try to introduce the ortolan bunting again (such an event is, of course, highly unlikely given conservation and biosecurity concerns), what would we predict would happen? Even better, given what we know already, what features of such a reintroduction (e.g., numbers of birds, season and location of release) would we likely think crucial to the long-term success?

In fact, even the most cursory of analyses suggests likely factors in determining the success or otherwise of introductions. For example, the number of founders appears, not surprisingly, to be highly relevant, although the strength of these numbers may be confounded by counting birds introduced after a species had become established (Moulton et al., 2011; but see Pipek et al., 2015). Nevertheless, over 600 yellowhammers were released on at least a dozen occasions at numerous sites around the country (Moulton et al., 2011; Pipek et al., 2015). Somewhere between 11 and 29 cirl buntings are known to have been introduced, at just two locations (Thomson, 1922; Moulton et al., 2011). The numbers of the other two species were smaller: six ortolan buntings and up to 11 common reed buntings were released on just one and two occasions each (Thomson, 1922; Moulton et al., 2011).

More sophisticated investigations give yet further insight. For example, the yellowhammer occurs at significantly greater densities in New Zealand compared to Britain, but this difference is not explained by differences in habitat availability, the abundance of invertebrates as food or levels of nest predation (Macleod et al., 2005). Ruling out these factors suggests that they may have been less important in the success of yellowhammers in New Zealand. Instead, Macleod et al. (2005) hypothesize that greater austral winter survival in the more benign New Zealand climate may be key. This factor may be critical for a second of these introduced bird species: changed agricultural practices that reduced winter survival of the cirl bunting are thought to explain its twentieth-century decline in England (Jeffs & Evans, 2004). If improved survival over winter is so important in the population dynamics of *Emberiza* buntings, it may also explain the differences in species-level success in New Zealand. Is it coincidental that the winter distribution of yellowhammers in its native range is the most northerly of the four species (del Hoyo, 2020), suggesting that it is the most resistant to cold?

At this point, the reader will realize that what is going on here is standard science: the bringing together of evidence that rules out some hypotheses and suggests others. In following this route, we simultaneously refine our ability to rule out some of the myriad of potentially causative factors that we previously lumped together under the label "contingency." Nevertheless, although such advances improve our explanatory power, they are not really what I mean when I make a plea to take history as contingency seriously.

3.3 Competitive Feeding Fish

Let us return to the motivating example of Chapter 1 (section 1.2), the question of how a group of differentially competitive fish should distribute themselves over two sites with different rates of food supply. I argued that the historical approach to solving this ICDD problem (Spencer et al., 1995) illustrated the flavor of contingency because the site choice of the newly arriving fish depends on the sites chosen previously by the fish already in the system.

There are two features of my use of an historical approach in this case that are worth noting here. First, each step in the process of getting to the final equilibrium depends on the sum total of what has happened before. It is the current distribution of fish that determines what happens next; the way in which we got to this point no longer matters. In a peculiar way, it is almost as if history doesn't matter. As an example, if, early on in the history, there are just three fish in the system, a good and a poor competitor at the better site and a poor competitor at the poor site, the choice of the incoming (fourth) fish is not dependent on whether the good or the poor competitor arrived first at the better site. Rather, the decision of the new arrival depends on the current 3:1 ratio of competitive abilities and the 2:1 ratio of food availability.

Such a memory-less property—what mathematicians and computer scientists call Markovian, dubbing such systems Markov processes—need not hold in all cases of contingency, however. For example, if the newly arriving fish had observed its predecessor arrive and this latter fish's choice of site somehow affected the latest arrival's decision (perhaps because of some social interactions or a perception of safety in numbers), the process would not be Markovian. Asking ourselves whether our system is Markovian, however, is useful. Not only does it increase our understanding of the system, but there are a number of ways some non-Markovian systems can be reparameterized so that they are Markovian. In some cases, too, the behavior of the system may be very close to Markovian, which then is a suitable approximation.

Box 3.1 Using the Markov Property of the ICDD to Find the Exact Distribution of Solutions

As described in section 3.3, the assumptions underlying the ICDD include the Markovian property that, in deciding which site to choose, newly arriving fish take account only of the current distribution of fish and the ratio of food available at the two sites. To be precise, the nth individual arriving in the system calculates its potential marginal gain at site i ($i = 1, 2$),

$$g_i = r_i / \left(c_n + \sum_{j=1}^{n-1} \delta_{ij} c_j \right)$$

where r_i is the food availability at site i ($r_i = 1$ or 2 in our numerical example), c_j is the competitive ability of individual j ($c_j = 1$ or 2 in our numerical example) and $\delta_{ij} = 1$ if individual j is at site i and 0 if not (Spencer et al., 1995). The individual then goes to site 1 if $g_1 > g_2$ and vice versa; if $g_1 = g_2$ either site is chosen with probability ½.

Now, let us assume that site 1 is the better site (i.e., $r_1 = 2$, $r_2 = 1$) and consider how the system progresses from the start, until all 12 fish, six good and six poor competitors, have been added. If the first fish is a good competitor (which occurs with probability ½) it calculates $g_1 = 2/(2 + 0) = 1$ and $g_2 = 1/(2 + 0) = ½$ and hence chooses site 1. A poor competitor (also arriving first with probability ½) also picks site 1, since, for it, $g_1 = 2$ and $g_2 = 1$. (It is fairly obvious that the first fish will pick the better site, of course!) So, after the first fish has decided, we have two possible distributions, each with probability ½:

and

Now, the second fish arriving in the scenario on the left calculates $g_1 = 2/(2 + 2) = ½$ and $g_2 = 1/(2 + 0) = ½$ if it is a good competitor, which occurs with probability 5/11, and chooses either site with equal probability. A poor competitor arriving with probability 6/11, however, calculates $g_1 = 2/(1 + 2) = ⅓$ and $g_2 = 1/(1 + 0) = 1$ and picks site 2. In the right-hand case, the good competitor (probability 6/11) calculates $g_1 = 2/(2 + 1) = ⅔$ and $g_2 = 1/(2 + 0) = ½$, thus picking site 1; the poor competitor (probability 5/11) calculates $g_1 = 2/(1 + 1) = 1$ and $g_2 = 1/(1 + 0) = 1$, thus picking either site with probability ½. These calculations illustrate a more general point made by Netz et al. (2023) that good competitors prefer better sites more than poor competitors do.

Putting all this together, we find we have six possible distributions, with associated probabilities as follows:

Box 3.1 *Continued*

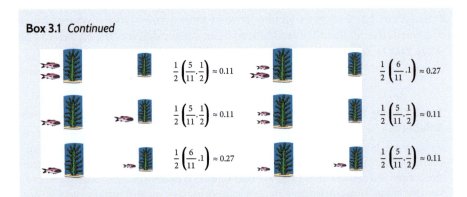

We could continue in this manner (called a "decision tree") for the next 10 steps, although the calculations become tedious. A series of contingent events can often be depicted very clearly in a decision tree.

Second, the Markovian nature of our historical model suggests that simulation need not be necessary. Instead, we could use some elementary probability theory to work out the exact probabilities of the different ICDD solutions (see Box 3.1). By looking at the distribution of fish already in the system, we can work out the probability that the newly arriving fish will choose one site or the other precisely. Combining these probabilities would give us the exact distribution of solutions to this ICDD.

The critical point here is that this form of Markovian contingency allows clear predictions (and hence generalizations). We need not give up when we realize that our biological questions contain some element of contingency.

It is also clear that the details of the historical dimension really matter. For example, it may be that the foragers could switch sites, perhaps after each arrival, as in Goss-Custard et al. (1995) and Houston and Lang (1998); see also Ruxton and Humphries (1999). These researchers found that poorer competitors were more likely to switch sites in such a model. Alternatively, foragers could reassess which site was best after all had arrived. In a model with constraints on how well the fish could perceive differences between sites, Spencer et al. (1995) found that these constraints interacted with the competitive differences to result in further undermatching with respect to the IFD or the ICDD.

In passing, I think it is worth observing that these models, whether historical or not, are examples of what Servedio et al. (2014) called "proof-of-concept" models. As I noted in Chapter 1, they are not designed to interface directly with data about the distribution of foragers in the real world (or even an experiment). Rather, they are a test of the correctness of the verbal hypothesis and assumptions concerning the way foragers select sites. Mathematical modeling has the virtue of making assumptions clear (or at least clearer); I find it interesting, though, that the ahistorical equilibrium assumption in the original versions of the IFD and ICDD was not immediately apparent.

The flavor of contingency clearly has some similarities here to that of construction (see Chapters 8 and 9): the models could be viewed as constructing the foragers' distribution over time. It seems to me, however, that the questions being asked are crucially different. Here, with contingency, these questions hinge on the way in which events (the choices of which site each fish goes to) depend on the current state of the system. The matter of interest is which of the many possible end results (equilibria in many cases) are realized.

In construction, the focus is on building up of genetic or community structure over time, not usually which alleles of species successfully invade. And the final result is not usually an equilibrium, since alleles and species can be continually replaced.

3.4 Wright's Shifting-Balance Theory

Sewall Wright's metaphor of the adaptive landscape must be one of the most influential in population genetics. Proposed early on in the development of the "Modern Synthesis," variations on this portrayal of the fitness of a population as a function of its composition (e.g., frequency of different phenotypes or genotypes; Wright, 1932) can be found in thousands of research papers today (see Fig. 3.2). The adaptive landscape was central to Wright's ideas about how adaptation proceeds, what he called his "shifting-balance theory" (Wright, 1929, 1931, 1932)[1]. A synthesis of the effects of genetic drift and selection acting on a subdivided population whose subpopulations (or demes) are nevertheless connected by some level of migration, the theory was proposed by Wright as a solution to the problem of adaptation. Under shifting balance, Wright held that the totality of demes could explore the adaptive landscape more efficiently than a single large population (see Box 3.2).

So why is the shifting-balance theory an example of history as contingency? In short, because the path to improved adaptation depends on which demes have shifted to the higher adaptive peak. Demes in closer geographical proximity are more likely to be connected by migration. Hence, demes that have shifted already are likely to be geographically closer to each other, as are those next in line to shift. The process of adaptation may proceed along many different routes, even if the endpoint is the same.

In addition, if there are several higher peaks nearby on the adaptive landscape, different peak shifts may occur, depending on which peak is colonized first in Phase 2 of the theory. Moreover, if that peak succeeds in attracting the rest of the population, the genetic constitution of the meta-population will be altered. Since different peaks reflect different genotypic combinations, the likelihood of subsequent peak shifts will also change.

---

[1] The naming of the shifting-balance theory, though, occurred much later. So far as I have been able to ascertain, Wright did not use the term until more than 30 years later (Wright, 1965).

36 • *Beyond Equilibria*

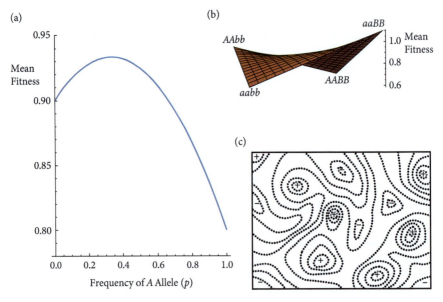

Fig. 3.2 Adaptive landscapes. (a) Population mean fitness for the model of heterozygote advantage in Box 2.2. Note that the fitness peak occurs at the allele-frequency equilibrium of $p = ⅓$. (b) Population mean fitness for a two-locus, two-allele model of selection. Note that, for this particular model, the global peak occurs for a population fixed for the a and B alleles. A population with high frequencies of the alternative A and b alleles, however, will be driven by selection towards fixation of those alleles, which corresponds to a local fitness peak. An increase in adaptation for this latter population requires that the population somehow crosses the adaptive valley (the saddle) to the higher peak. (c) Wright's original diagram drew the adaptive landscape as a topographical map, where the axes represent frequencies of genotypic combinations, so that nearby points on the landscape are similar genetically. The dotted curves—analogous to contour lines—connect points of equal fitness, with peaks indicated by plus signs.

(c) Reproduced from Wright (1932) courtesy of the Brooklyn Botanic Garden.

Box 3.2 The Shifting-Balance Theory

The shifting-balance theory envisages adaptation occurring as an increase in the mean fitness of a population. On an adaptive landscape, such an increase corresponds to a population moving from one adaptive peak to a higher one. Wright described the process as having three stages or "phases" as he called them.

Phase 1: Demes explore the adaptive landscape. Selection will tend to push these demes back towards the local fitness optimum, a peak on the adaptive landscape. This process will be more efficient in larger demes. Genetic drift, however, which is accentuated in smaller demes, will allow demes to wander away from an adaptive peak. Occasionally, a deme that has wandered far from its previous adaptive peak will cross an adaptive valley.

Phase 2: Selection within this latter deme drives it towards the new local adaptive peak. This deme has undergone a peak shift.

Box 3.2 *Continued*

Phase 3: Inter-group selection then drives the remaining populations of the species to the new peak. Demes at higher adaptive peaks will, by virtue of their greater fitness, produce more offspring than demes at lower peaks. Migration from these fitter demes then drags less well adapted demes towards the new peak. Note that this demic movement is on the adaptive landscape; it is not geographical. (Nevertheless, migration is more likely between demes that are geographically close.) These latter demes may then cross the adaptive valley and are then pushed towards the new peak by selection. At this point, the whole population has undergone a peak shift and adaptation to the environment has improved.

In Wright's view, population subdivision was especially conducive to adaptation. Compared to a larger unstructured population, the greater genetic changes wrought by drift in the smaller subdivided demes enabled exploration of a larger part of the adaptive landscape. Large panmictic populations would, according to Wright, be prone to remain stuck at lower peaks, held there by more efficient selection.

The shifting-balance theory has provoked significant modeling research, especially in the last 30 years. Much of this effort has gone into investigating the third phase, which was controversial because it was unclear whether or not the migrational pull of the single population at the new peak was sufficiently strong to counter the effects of back migration into that peak from the potentially large number of unshifted populations (Whitlock & Phillips, 2000).

Finally, environmental changes change the shape of the adaptive landscape, and so how these changes take place will affect the way in which any peak shifts might occur. In summary, then, the process of adaptation of the population as a whole is strongly contingent on what has already happened: the demes that have already shifted (both geographically and genotypically) as well as any environmental perturbations.

3.5 Different Outcomes in Competitive Communities

My third example of contingency in eco-evolutionary theory comes from the investigation of the mathematical properties of a model of ecological community structure, the well-known Lotka-Volterra equations. This model comes in a number of different versions (see Chapter 9 for a slightly different formulation) but Gilpin and Case (1976) examined simulations of the equation for the rate of change in the normalized population density of species i:

$$\frac{dN_i}{dt} = N_i \left(1 - \sum_j a_{ij} N_j \right).$$

Here, N_i is the density of species i ($i = 1, 2, \ldots n$; $0 \leq N_i \leq 1$) and the $a_{ij} \geq 0$ (with $a_{ii} = 1$) are competition parameters. The greater an a_{ij} value, the more species j impacts the population size of species i.

For a given matrix of randomly selected a_{ij} ($i \neq j$), Gilpin and Case (1976) iterated the above equations starting with different (random and constrained) N_i values. The use of simulations meant that the properties of the model with different parameters (the a_{ij} values) and different initial variable values (the N_i values at the start of each simulation) could be systematically explored. In their iterations, they deleted the equations and parameters for species if they became extinct, until an equilibrium was reached.

Different starting points (initial values of the N_i) iterated to different equilibria, even when the same matrix of competition parameters was used: the endpoints were contingent on the starting values. Moreover, initially larger systems had, on average, more separate equilibria (see Fig. 3.3).

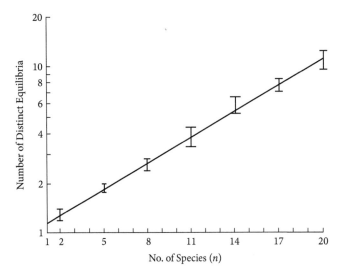

Fig. 3.3 The mean number of different final equilibria as a function of the initial number of species in the computer simulations of Gilpin and Case (1976). The bars indicate ±1 standard error. Redrawn with permission after Gilpin and Case (1976).

It is worth noting that the Lotka-Volterra equations above are completely deterministic; the only stochasticity was in the initial population densities. By contrast, my first example of contingency (of the competitive foraging fish) had randomness at every time step, as the order of the individual fish arriving in the system was random.

But why does all of this matter? As the authors so eloquently put it as they began their paper:

> Ecologists typically explain the absence of a species in an area by searching for ultimate climatic or physical factors that limit its distribution or favour its competitors. The possibility that the system is not in equilibrium—that colonisation has not occurred—is not normally considered. Even less appreciated is that there may be a number of alternative stable equilibrium communities which may develop in a given area and that the final outcome

may depend solely or partially on historical factors such as the sequence and numbers in which each species colonises.

Clearly, in their opinion, history is critical to the understanding of the makeup of ecological communities. Even better, this quotation identifies as distinct two of my flavors of history: turnover (the system possibly not being at equilibrium) and contingency (the dependency on the order of events and the state of the system). Of course, both aspects of this quotation substantiate my disclaimer in Chapter 1 that my emphasis on the importance of history and its presence in different flavors is not entirely novel.

3.6 Contingency in Perspective

The dependence of so many events on the history of a biological system means that contingency is undoubtedly one of the most important and widely recognized of my historical flavors. Although it might seem that the particularities of contingent events preclude scientific generalizing, it seems to me that standard methods can certainly make progress. Moreover, a modeling approach that replicates the events and examines a range of possible outcomes allows systematic investigation of contingency and there is a plethora of examples in the ecological and evolutionary literature. Indeed, replication can even be achieved in some observational and experimental studies, thus revealing the role of contingency empirically. In some cases, the contingent nature of the system can be depicted in a decision tree, which can show all possible outcomes and their relative likelihoods.

References

Del Hoyo J. (ed.). 2020. *All the Birds of the World*. Barcelona: Lynx Edicions.
Gill B. J., Bell B. D., Chambers G. K., Medway D. G., Palma R. L., Scofield R. P., Tennyson A. J. D., Worthy T. H. 2010. *Checklist of the Birds of New Zealand, Norfolk and Macquarie Islands, and the Ross Dependency, Antarctica*. Wellington: Te Papa Press.
Gilpin M., Case T. 1976. Multiple domains of attraction in competition communities. *Nature* 261:40–42.
Goss-Custard J. D., Caldow R. W. G., Clarke R. T., Durell S. E. A. le V. dit, Sutherland W. J. 1995. Deriving population parameters from individual variations in foraging behaviour. I. Empirical game theory distribution model of oystercatchers *Haematopus ostralegus* feeding on mussels *Mytilus edulis*. *Journal of Animal Ecology* 64:265–276.
Houston A. I., Lang A. 1998. The ideal free distribution with unequal competitors: The effects of modelling methods. *Animal Behaviour* 56:243–251.
Jeffs C., Evans A. 2004. Cirl bunting: The road to recovery. *Biologist* 51:189–193.
Koch E. L., Ravinet M., Westram A. M., Johannesson K., Butlin R. K. 2022. Genetic architecture of repeated phenotypic divergence in *Littorina saxatilis* ecotype evolution. *Evolution* 76:2332–2346.

Macleod C. J., Wratten S. D., Duncan R. P., Parish D. M. B., Hubbard S. F. 2005. Can increased niche opportunities and release from enemies explain the success of introduced Yellowhammer populations in New Zealand? *Ibis* 147:598–607.

Moulton M. P., Cropper W. P., Avery M. L. 2011. A reassessment of the role of propagule pressure in influencing fates of passerine introductions to New Zealand. *Biodiversity & Conservation* 20:607–623.

Moulton M. P., Cropper W. P. Avery M. L., Avery M. L. 2012. Historical records of passerine introductions to New Zealand fail to support the propagule pressure hypothesis. *Biodiversity & Conservation* 21:297–307.

Netz C., Ramesh A., Weissing F. J. 2023. Ideal free distribution of unequal competitors: Spatial assortment and evolutionary diversification of competitive ability. *Animal Behaviour* 201:13–21.

Pipek P., Pyšek P., Blackburn T. M. 2015. How the Yellowhammer became a Kiwi: The history of an alien bird invasion revealed. *NeoBiota* 24:1–31.

Ruxton G. D., Humphries S. 1999. Multiple ideal free distributions of unequal competitors. *Evolutionary Ecology Research* 1:635–640.

Servedio M. R., Brandvain Y., Dhole S., Fitzpatrick C. L., Goldberg E. E., Stern C. A., Van Cleve J., Yeh D. J. 2014. Not just a theory—The utility of mathematical models in evolutionary biology. *PLoS Biology* 12:e1002017.

Spencer H. G., Kennedy M., Gray R. D. 1995. Patch choice with competitive asymmetries and perceptual limits: The importance of history. *Animal Behaviour* 50:497–508.

Thomson G. M. 1922. *The Naturalisation of Animals and Plants in New Zealand* (Cambridge Library Collection—Zoology). Cambridge, UK: Cambridge University Press.

Whitlock M. C., Phillips P. C. 2000. The exquisite corpse: A shifting view of the shifting balance. *Trends in Ecology & Evolution* 15:347–348.

Wright S. 1929. Evolution in a Mendelian population. *Anatomical Record* 44:287.

Wright S. 1931. Evolution in Mendelian populations. *Genetics* 16:97–159.

Wright S. 1932. The roles of mutation, inbreeding, crossbreeding, and selection in evolution. *Proceedings of the Sixth International Congress on Genetics* 1:355–366.

Wright S. 1965. Factor interaction and linkage in evolution. *Proceedings of the Royal Society of London B* 162:80–104.

4

Constraint

4.1 History as Constraint

I use the term "constraint" as a sort of extreme contingency, describing the situation in which particular future events are ruled out (or become certain), rather than, as with contingency, made more or less likely. A constraint arises because of what has already transpired, which determines what can or cannot happen next. Wright's shifting-balance theory (see section 3.4) actually provides us with our first example of history as constraint. One peak shift may render some other shifts impossible if it removes certain genetic variants from the meta-population.

The concept of constraint is common in developmental biology. Developmental constraints restrict the possible ways in which a character (or suite of characters) can develop. Such restrictions arise because of irreversible changes that commit subsequent development to particular pathways. Conrad Waddington's (1942) well-known concept of canalization (Fig. 4.1) is an example. The model was originally proposed as a way to envisage developmental homeostasis, the rather limited phenotypic variation shown by adult organisms given the possibility of significant genotypic variation: the ball rolling towards us can take only certain paths. But these developmental pathways have bifurcations, points at which development could go in different directions and some sort of "developmental decision" must be made. Those (obviously unconscious) choices have consequences later on, when only certain developmental pathways are then possible.

Another example of constraint might be some of Diamond's (1975) "assembly rules" that derive from a consideration of the way potential species in a community compete. For example, Diamond (1975) held that the occurrence of different species from a guild of honeyeaters and sunbirds on the Bismarck Islands off New Guinea was decidedly non-random. The presence of a pair of species on a particular island could not always be predicted by assuming their occurrences were independent. Rather, their distributions were driven by strong dietary competition, combined with aggressive behavior. As a consequence, certain pairs of species—such as the sunbird, *Cinnyris frenatus* (then identified as *Nectarinia jugularis*) and the honeyeater *Myzomela sclateri*—were "forbidden," never observed. Such exclusions are a form of constraint. Other assembly rules that result in less extreme outcomes (e.g., non-zero, albeit lower, frequencies of co-occurrence than expected) might be considered a form of contingency.

Beyond Equilibria. Hamish G. Spencer, Oxford University Press. © Hamish G. Spencer (2025).
DOI: 10.1093/oso/9780192858993.003.0004

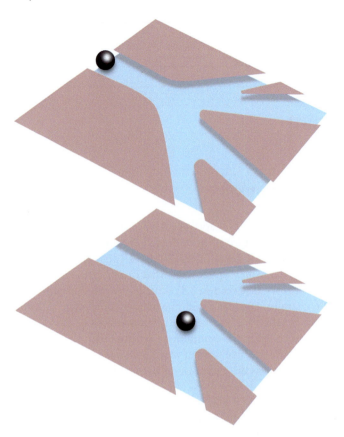

Fig. 4.1 In the upper figure, the ball represents the organism early in development. The path it takes as it rolls to the right (developing as it goes) involves certain "decisions" about the path to be taken, which constrain the subsequent developmental decisions. In the lower figure, a decision to take the right-hand path at the first choice has already been made, thus constraining the second decision, which is about to be made.

Inspired by Waddington (1957).

Let us, however, look at a different example in detail, that of the search for the smallest possible genome.

4.2 The Minimal Gene Set for a Free-living Cell

Microbiologists have long been interested in working out which genes are indispensable in the genome of the simplest free-living organisms (Koonin, 2000). This "minimal gene set" should be capable of allowing a single-celled microbe to survive and reproduce under ideal conditions (i.e., a benign environment, supplying all essential nutrients with no predators, competitors or abiotic stressors). Such information is also needed in producing cells with synthetic genomes, which can help us understand the fundamental properties of cell function (Pelletier et al., 2021). A related but slightly different issue (see section 4.3) is the minimal number of genes resulting from the genome reduction that has occurred in some eubacterial intracellular bacteria, such as the *Buchnera* endosymbionts of aphids (Gil et al., 2002). This number might be expected to be smaller than the number required by a free-living cell because some gene products can be acquired by the bacterium from its host.

The most straightforward approach to these questions is simply to look at the genomes of existing microbes, examining in detail those that have the fewest genes. Mycoplasmas, for instance, have long been thought to have close to a minimal gene set. A human-urogenital pathogen, *Mycoplasma genitalium*, is the current record holder for the smallest known genome of any naturally occurring microorganism able to be maintained as a pure culture, with 482 known protein-coding sequences, three phosphate-transporter genes, and 43 RNA-coding genes (Glass et al., 2006). So, is 528 (= 482 + 3 + 43) the answer? The obvious problem with this number is that we have no certainty about whether further genes could be lost without compromising viability. Nevertheless, 528 is a useful upper bound.

Adding a modicum of a temporal dimension to the investigation—not exactly history, I admit, but a step in the right direction—can directly address this last problem. Glass et al. (2006) examined the effect of the insertion of transposons that would be expected to disrupt gene function in *M. genitalium*. Effectively, they were asking if further loss of individual genes was possible. Their results showed that 100 different protein-coding genes could be disrupted and yet clonal populations of these single mutants remained viable. Subtracting these 100 "putatively nonessential" genes from the total of 482 implied that 382 protein-coding sequences were essential. Consideration of which gene families had been disrupted meant they increased this number to 387, with the addition of five further genes they considered indispensable.

Interestingly, a knockout study on the Gram-positive bacterium *Bacillus subtilis*, which has a considerably larger genome of ~4100 genes, gave a much lower number. After examining clones with individual genes inactivated, Kobayashi et al. (2003) concluded that 271 genes are necessary for growth. This figure is a surprisingly low fraction (6.6%) of the genes in *B. subtilis*, as well as being considerably less than 387.

A different way of attacking this issue would use a comparative approach, looking for overlap in the genomes of distantly related microbes. The vast evolutionary distance between the comparison species means that they are less likely to share essential genes simply because of common descent. Good candidates include disease-causing bacteria that can obtain most if not all of their nutrients from their environment, especially the organisms they infect, and which have consequently been able over evolutionary time to jettison unneeded parts of their genomes. In one of the first studies, Mushegian and Koonin (1996) compared *M. genitalium* and the human-pathogenic bacterium *Haemophilus influenzae*, which have significantly smaller genomes than their respective presumptive ancestors. *H. influenzae* has a larger genome than *M. genitalium*, with ~1700 genes (Iksander et al., 2017). The most recent common ancestor of the two species is thought to have lived over 3 billion years ago.

Mushegian and Koonin (1996) found that the two species shared 240 genes coding for similar amino acid sequences, producing proteins with comparable functions. To this number, they added a further 22 genes involved in critical biochemical reactions but which did not show similar sequences in the two species. These additions filled gaps in essential metabolic pathways. Finally, they considered that six further genes were functionally redundant or specific to being a parasite, and removed them from the list. The resultant total of 256 was considered an upper bound on the size of the minimal gene set, a figure again rather lower than 387.

But as Koonin (2000) pointed out, pursuing this path by including more species gives an answer that is too small: there would appear to be only ~80 universal orthologs (see Jensen, 2001 for clarification of this terminology). History really matters here: clearly different lineages have been able to satisfy various functional requirements with non-orthologous genes (Moran, 2002). Indeed, closer scrutiny of genes in small genomes can pay dividends: by including more species in their comparison as well as examining the functions of different genes more carefully, Gil et al. (2004) came up with a lower estimate of 206 essential protein-coding genes.

Why are these numbers all over the shop? One answer is to invoke the details of the system being examined: some genes may be essential in some circumstances but not in others (e.g., depending on exactly what nutrients the environment supplies, or because the clones containing disruptive mutants cannot survive in the lab but would do so in the natural habitat). But surely the major reason is to do with constraints, which naturally arise from the ubiquitous interactions among gene products. Even the adjusted figure of 387 from Glass et al. (2006) may be too low, for example, if one of Gene A and Gene B can be lost, but not both. Once certain genes have been dispensed with, other genes, previously simply useful, may become essential. Their subsequent loss would be lethal. The method of looking at the effects of single-gene knockouts is likely to err in counting both as unneeded.

What is really needed in understanding the process of genome reduction is an historical approach, one that seriously incorporates constraint. Interestingly, Waddington's (1957) figure (Fig. 4.1) is also a fitting visual metaphor here. As the ball rolls down the slope, each bifurcation represents the loss or otherwise of a particular gene. If that gene is lost, the set of potential future losses is irrevocably changed. So, what would an historical approach look like? Although undoubtedly technically challenging, a first experimental step would be to look at double mutants, in which two genes have been knocked out. My expectation is that, whatever the system, a good number of genes classified as nonessential in single-mutant studies may turn out to be essential in these hypothetical expanded studies and, thus, the estimates of minimal genome size would increase. More fundamentally, it is not really the size of the genome that counts; it is the content: it is which genes are present (or absent) that really matters.

4.3 The Minimal Gene Set for an Endosymbiont

Interestingly, a sort of natural experiment that corresponds to repeated use of the Waddington landscape has played out in *Buchnera*. These maternally transmitted bacteria live in the cytoplasm of specialized cells of their aphid hosts and, by supplying essential amino acids, allow many aphid species to survive on what would otherwise be a nutrient-deficient diet. This obligate association is truly ancient, dating back at least 100 million years (Martinez-Torres et al., 2001; van Ham et al., 2003), and the phylogenies of the aphids and their *Buchnera* passengers show remarkable congruence (Nováková et al., 2013).

An analysis of 39 different *Buchnera* strains sampled from across the aphid (Aphididae) phylogeny showed considerable differences—between 354 and 587—in the number of protein-coding genes (Chong et al., 2019). Each of these strains corresponds to a ball at the front edge of the Waddington landscape; how far back they started depends on when they separated evolutionarily from other strains. But once on the way towards us, each ball/strain is on an independent path from every other. In this scenario, we would expect balls that had been together more recently (which represent more closely related strains) to have lost more genes in common and thus, because of these resulting constraints, to have more similar-sized genomes. And that is exactly what we find: the strains commensal with aphids in the same subfamily or tribe have comparable numbers of protein-coding genes and genome sizes (Chong et al., 2019; Fig. 4.2).

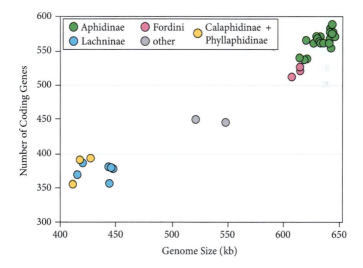

Fig. 4.2 The number of protein-coding genes and the genome size of *Buchnera* strains symbiotic with 39 aphid species from different parts of the Aphididae phylogeny. The clustering of points of the same color shows that strains of *Buchnera* commensal with more closely related aphids have similar genome sizes and numbers of protein-coding genes.

Redrawn from Chong et al. (2019), with permission.

More critically, we would expect strains from closely related aphids to have lost similar sets (not just similar numbers) of genes. Again, this expectation is fulfilled: genes involved in cell-wall production, for example, have been preferentially lost outside the Aphidinae clade (Chong et al., 2019). Nevertheless, even this prediction is rather weak, since the loss of sets of genes may simply reflect a single set of losses in the ancestor of a clade. In other words, we are seeing the effects of a phylogenetic constraint: related strains share similar losses precisely because they are related.

To see the true effects of historical constraint we need to examine genes that have been lost on separate occasions in more than one lineage. Chong et al. (2019) found 18 genes that had been lost independently six or more times during the evolution of *Buchnera*. Mapping these losses onto the phylogeny and seeing if different genes were lost together more or less often than expected at random is a statistical issue beyond this analysis (not least because the phylogeny should be re-estimated with these genes omitted). Nevertheless, it is telling that, as Chong et al. (2019) noted, these 18 genes were often related functionally: four genes involved in stress response were repeatedly lost, whereas nearly all genes underlying amino-acid biosynthesis were always retained. Clearly, there is some constraint involved in these evolutionary pathways.

A second aspect of this study provides further evidence of the importance of constraint. Aphids in the Lachninae have a second endosymbiont, a strain of an unrelated bacterium, *Serratia symbiotica* (Lamelas et al., 2011). One lachnine aphid, *Cinara cedri*, hosts a strain of *Buchnera* that cannot synthesize the amino acid tryptophan, normally provided to aphids by *Buchnera*. In *C. cedri* tryptophan is produced by *Serratia*. The presence of a second endosymbiont has released *Buchnera* from the constraint of synthesizing tryptophan and it has responded by eliminating the responsible gene. Indeed, this strain has the smallest known *Buchnera* genome. Lamelas et al. (2011) speculated that the *C. cedri* strain of *S. symbiotica*, which has a notably smaller genome than others of its species, is on an evolutionary path leading to obligate endosymbiosis. Most importantly, however, they concluded that "there are several cases of metabolic complementation giving functional stability to the whole consortium [of aphid and bacteria]". Historical constraint, in other words (Douglas, 2016).

Maniloff (1996) pointed out that there are, in fact, two distinct evolutionary routes to minimal genomes, which he labeled "top down" and "bottom up." The former route is a reduction over evolutionary time and underlies the logic outlined above, whereas the latter is an increase, thought to be taken by the first living organisms, the earliest of which presumably had very small genomes. This focus on the evolutionary pathway—the history of the system—is, I think, most illuminating. Maniloff (1996) noted that the metaphor of evolution as an engineer, working on already well-adapted genomes, applies to the top-down reduction. By contrast, viewing evolution as a tinkerer, making do with whatever materials are at hand, is relevant to the bottom-up increase to a fully functional genome. These alternatives imply that very different genes are likely to be found in minimal genomes. Both pathways are subject to constraints, however.

4.4 A Metazoan Example

Genome reduction has occurred in metazoans as well, again the result of the evolution of symbiosis. The mite *Demodex folliculorum* spends its whole 16-day life cycle on the human body, on our skin and in our hair follicles, especially on our faces (see Fig. 4.3). At least 90% of us (and maybe everyone) carry them from soon after birth (Foley et al., 2021). During the night males emerge from our hair follicles and crawl over our skin in search of females, who remain inside their own follicles (Pormann et al., 2021). Smith et al. (2022) showed that this species is usually passed on maternally, possibly during infant breastfeeding. Although higher densities of *D. folliculorum* are associated with some skin diseases (Pormann et al., 2021), it appears that the relationship between humans and this mite is usually commensal (Foley et al., 2021).

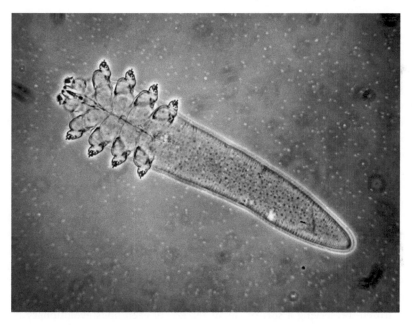

Fig. 4.3 Photo of the human commensal skin mite, *Demodex folliculorum*, showing its transparent, worm-like body and four pairs of legs. Total length is ~0.35 mm.
Photo: © Henry Perry, MD, with permission.

This mite's highly specialized habitat means the species has no predators or competitors. The resultant relaxed selection, in conjunction with high levels of inbreeding and pronounced genetic drift, arising from the strong population-bottleneck effect during maternal transmission, has led to a host of linked changes in the species' genome, body structure and behavior. The genome is among the smallest known in the invertebrates (< 52 Mbp and just 9707 coding genes; Smith et al., 2022). From a phylogenetic comparison with other mite species, Smith et al. (2022) argued that *D. folliculorum* has experienced relatively recent but extensive gene loss across a large number of gene families. A number of DNA-repair genes, for example, appear to be absent, and genes involved in protection against UV light, unnecessary in a nocturnal animal, have been lost.

The mites cannot produce the hormone melatonin and appear to depend on consuming the human form for its essential roles in mite reproduction and circadian activity pattern. All of these changes lock the species into a closer and closer relationship with its human host, on which it has become completely dependent, further constraining its future evolution. The series of evolutionary constraints has had critical consequences. Indeed, Smith et al. (2022) speculated that aspects of this evolutionary history, notably the loss of DNA-repair genes, doom the species to extinction as deleterious mutations accumulate. Paradoxically, this increased extinction risk has evolved at the same time as the mite's reduced pathogenicity.

4.5 Gain of Function in a Microbial Experiment

Historical constraint naturally arises during evolutionary loss of function. But this next example illustrates that it can also occur in gain-of-function scenarios. Although described by the authors as an example of historical contingency, I prefer to think of it as historical constraint because it seems that the end result of the evolutionary process would have been impossible (or nearly so) without a series of critical steps along the way. As I noted at the start of the chapter, I view constraint as an extreme form of contingency, in which, rather than the likelihood of different outcomes being altered, an outcome is completely eliminated (when the constraint is imposed) or (when the constraint is removed) the outcome becomes possible.

Almost 40 years ago, evolutionary biologist Richard Lenski initiated the "long-term evolution experiment" (LTEE), a remarkably simple yet highly influential study using the asexual bacterium, *Escherichia coli* (Lenski et al., 1991). He founded 12 replicate lines that have been maintained, isolated from each other, ever since, and tracked the genetic and phenotypic changes that have evolved over time (see Fig. 4.4). Critically, every 500 generations (about every 75 days) a sample from each of the replicates was frozen, preserving a sort of "fossil record" of each line's evolutionary history. These samples could be revived for comparative purposes.

One of the most interesting changes was the novel ability, which arose in just a single line, to utilize citrate as an energy source in an aerobic environment (Blount, 2016). Although citrate had always been present as a food source in the LTEE and *E. coli* can grow on citrate under *anaerobic* conditions, this new capability took > 31,000 generations to appear. Clearly, this change was not a matter of a simple mutation.

To investigate how this capacity for aerobically metabolizing citrate evolved, the researchers performed re-runs of the LTEE initiated with samples from different times in the frozen "fossil record" of the citrate-feeding line. They then searched for any repetition of this evolutionary change and examined the DNA sequences of these samples and their descendants. They concluded that the most plausible sequence of events was the occurrence of at least two "potentiating mutations" that allowed the

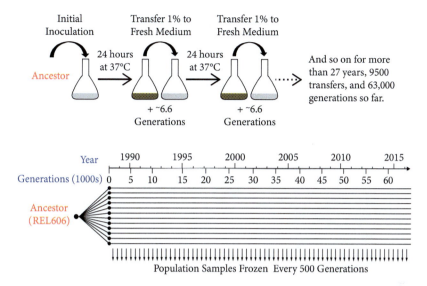

Fig. 4.4 Protocol for Lenski's long-term evolution experiment (LTEE) using *Escherichia coli*. From Blount (2016), with permission.

subsequent genetic switch for growing on citrate to be mutationally reachable (Blount, 2016). (Yet further mutations refined the initial poor ability to metabolize citrate aerobically, substantially improving the functional efficiency.) In other words, the context for the evolution of this change was all-important and, indeed, completely necessary. The absence of the potentiating mutations in the vast majority of the lines acts as a constraint, preventing the evolution of the ability of *E. coli* to grow on citrate in the presence of oxygen in nearly all circumstances.

Note that this study was able to provide its answer only because it had preserved its history in a way that allowed it to be carefully examined for the potentiating mutations. Taking history seriously clearly paid dividends and revealed the importance of constraint.

4.6 Adaptive Radiation in Scottish Sticklebacks

A recent study of three-spined sticklebacks (*Gasterosteus aculeatus*) on two neighboring islands off the northwestern coast of Scotland provides a further possible example of constraint (Begum et al., 2023). In several parts of the Northern Hemisphere over the last 10,000 years, populations of sticklebacks have evolved to live in freshwater and brackish habitats. These fish derive from anadromous ancestors, which live in the sea but migrate to brackish and freshwater habitats to breed. In many cases, this evolution has seen a substantial reduction in the size and number of various skeletal structures known as armor traits, which provide protection against predators, but may be expensive to produce, especially in nutrient-poor habitats (see Fig. 4.5). Such morphological changes in *Gasterosteus* have long been held up as examples of adaptive radiation (e.g., Schluter, 1993; Bell & Foster, 1994).

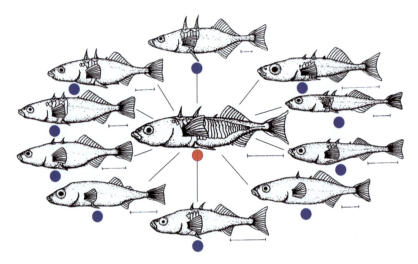

Fig. 4.5 Adaptive radiation in the three-spined stickleback (*Gasterosteus aculeatus*), with the ancestral diadromous form in the center (with the red dot) and derived freshwater and brackish forms (purple dots).
Adapted from Bell and Foster (1994), with permission.

Begum et al. (2023) explored why sticklebacks from North and South Uist show strikingly different levels of skeletal variation. Fish resident in interconnected freshwater and brackish lochs on North Uist exhibit some of the greatest diversity known in the species, whereas fish from the ecologically similar South Uist show much reduced, albeit measurable, variation. This difference is puzzling, since ancestral variation is probably shared and environmental selective pressures are likely to be similar. Indeed, Begum et al. (2023) showed that samples of anadromous fish collected from North and South Uist did not differ significantly from each other in their armor traits, in keeping with the assumption that ancestral variation was shared.

As Begum et al. (2023) point out, explanations for the sort of adaptive radiation seen on North Uist emphasize the ecological opportunities provided by the diversity of under-utilized resources in newly colonized habitats lacking in competitors. But these ideas do not seem to account for the observations from the non-marine populations on South Uist. Begum et al. (2023) suggest that, even if there is an absence of competitors in the colonized environments, resources may still be limiting, with abiotic factors constraining any adaptive radiation.

Begum et al. (2023) showed that the skeletal variation on both islands was associated with variation in the pH of the different lochs, a result also found in a number of previous studies elsewhere. But variation in pH was greater on North Uist, with two of its lochs manifesting particularly low pH and containing unusually small fish that lacked certain armor features. The authors interpret these results as more consistent with an explanation of ecological constraint (due to pH) than one invoking ecological opportunity. The greater variation in the North is simply because the constraint is weaker.

This case is interesting because it relies on an inference about the morphology of plausible ancestors of the non-marine populations. The recency of much stickleback adaptive radiation in these habitats means that current anadromous populations provide excellent proxies for such ancestors. In addition, the replication of the colonization events provided by populations from the two islands, together with the mismatch in evolutionary outcomes, alerts us to the need for an explanation that covers high and low levels of diversification. Investigations that make use of an historical approach often require both information about original states and some degree of replication.

Some readers might object that my use of constraint here is a little different from that in the previous examples. Even if we agree that the restricted range of pH on South Uist does indeed constrain the morphological variation in the sticklebacks, this constraint might not seem to result from the evolutionary history of these fish. I beg to differ, however: the evolutionary history of the sticklebacks must be seen in context, and include changes in their environment. Such arguments concerning the inseparability of organism and environment have been given eloquently by Sultan (2015).

4.7 Constraint in Perspective

Historical constraint imposes restrictions on the possible ecological and evolutionary outcomes of a biological system. What may have been potential outcomes at the beginning are no longer possible as the system develops. Constraint differs from contingency in precluding certain outcomes; contingency merely alters the chances that these outcomes may occur.

Waddington's metaphor for developmental canalization of a ball rolling forward down a wrinkled slope is helpful in showing how different options are ruled out as history progresses. Constraint is particularly evident in the reduction of microbial and other genomes, but it is also present in the evolution of new functions, which cannot take place before various prerequisites have evolved.

References

Begum M., Nolan V., MacColl A. D. C. (2023). Ecological constraint, rather than opportunity, promotes adaptive radiation in three-spined stickleback (*Gasterosteus aculeatus*) on North Uist. *Ecology and Evolution* 13:e9716.

Bell M. A., Foster S. A. 1994. Introduction to the evolutionary biology of the threespine stickleback. Pages 1–27 in *The Evolutionary Biology of the Threespine Stickleback*, edited by M. A. Bell, S. A. Foster. Oxford, UK: Oxford University Press.

Blount Z. D. 2016. A case study in evolutionary contingency. Studies in History and Philosophy of Science Part C: *Studies in History and Philosophy of Biological and Biomedical Sciences* 58:82–92.

Chong R. A., Park H., Moran N. A. 2019. Genome evolution of the obligate endosymbiont *Buchnera aphidicola*. *Molecular Biology and Evolution* 36:1481–1489.

Diamond J. M. 1975. Assembly of species communities. Pages 342–444 in *Ecology and Evolution of Communities*, edited by M. L. Cody, J. M. Diamond. Cambridge, Mass.: Belknap Press of Harvard University Press.

Douglas A. E. 2016. How multi-partner endosymbiosis functions. *Nature Reviews Microbiology* 14:731–743.

Foley R., Kelly P., Gatault S., Powell F. 2021. *Demodex*: A skin resident in man and his best friend. *Journal of the European Academy of Dermatology and Venereology* 35:62–72.

Gil R., Sabater-Muñoz B., Latorre A., Silva F. J., Moya A. 2002. Extreme genome reduction in *Buchnera* spp.: Toward the minimal genome needed for symbiotic life. *Proceedings of the National Academy of Sciences, USA* 99:4454–4458.

Gil R., Silva F. J., Peretó J., Moya A. 2004. Determination of the core of a minimal bacterial gene set. *Microbiology & Molecular Biology Reviews* 68:518–537.

Glass J. I., Assad-Garcia N., Alperovich N., Yooseph S., Lewis M. R., Maruf M., Hutchison C. A. III, Smith H. O., Venter J. C. 2006. Essential genes of a minimal bacterium. *Proceedings of the National Academy of Sciences, USA* 103:425–430.

Iskander M., Hayden K., Van Domselaar G., Tsang R. 2017. First complete genome sequence of *Haemophilus influenzae* serotype a. *Genome Announcements* 5:e01506–16.

Jensen R. A. 2001. Orthologs and paralogs—we need to get it right. *Genome Biology* 2:interactions1002.1.

Kobayashi K., Ehrlich S. D., Albertini A., Amati G., Andersen K. K., Arnaud M., et al. 2003. Essential *Bacillus subtilis* genes. *Proceedings of the National Academy of Sciences, USA* 100:4678–4683.

Koonin E. V. 2000. How many genes can make a cell: The minimal-gene-set concept. *Annual Review of Genomics and Human Genetics* 1:99–116.

Lamelas A., Gosalbes M. J., Manzano-Marín A., Peretó J., Moya A., Latorre A. 2011. *Serratia symbiotica* from the aphid *Cinara cedri*: A missing link from facultative to obligate insect endosymbiont. *PLoS Genetics* 7:e1002357.

Lenski R. E., Rose M. R., Simpson S. C., Tadler S. C. 1991. Long-term experimental evolution in *Escherichia coli*. I. Adaptation and divergence during 2000 generations. *American Naturalist* 138:1315–1341.

Maniloff J. 1996. Commentary. The minimal cell genome: "On being the right size." *Proceedings of the National Academy of Sciences, USA* 93:10,004–10,006.

Martinez-Torres D., Buades C., Latorre A., Moya A. 2001. Molecular systematics of aphids and their primary endosymbionts. *Molecular Phylogenetics and Evolution* 20:437–449.

Moran N. A. 2002. Microbial minimalism: Genome reduction in bacterial pathogens. *Cell* 108:583–586.

Mushegian A. R., Koonin E. V. 1996. A minimal gene set for cellular life derived by comparison of complete bacterial genomes. *Proceedings of the National Academy of Sciences, USA* 93:10,268–10,273.

Nováková E., Hypša V., Klein J., Foottit R. G., von Dohlen C. D., Moran N. A. 2013. Reconstructing the phylogeny of aphids (Hemiptera: Aphididae) using DNA of the obligate symbiont *Buchnera aphidicola*. *Molecular Phylogenetics and Evolution* 68:42–54.

Pelletier J. F., Sun L., Wise K. S., Assad-Garcia N., Karas B. J., Deerinck T. J., Ellisman M. H., Mershin A., Gershenfeld N., Chuang R. Y., Glass J. I., Strychalski E. A. 2021. Genetic requirements for cell division in a genomically minimal cell. *Cell* 184:2430–2440.

Pormann A. N., Vieira L., Majolo F., Johann L., Silva G. L. 2021. *Demodex folliculorum* and *Demodex brevis* (Acari: Demodecidae) and their association with facial and non-facial pathologies. *International Journal of Acarology* 47:396–403.

Schluter D. 1993. Adaptive radiation in sticklebacks: Size, shape, and habitat use efficiency. *Ecology* 74:699–709.

Smith G., Manzano-Marín A., Reyes-Prieto M., Antunes C. S. R., Ashworth V., Goselle O. N., Jan A. A. A., Moya A., Latorre A., Perotti M. A., Braig H. R. 2022. Human follicular mites: Ectoparasites becoming symbionts. *Molecular Biology and Evolution* 39:msac125.

Sultan S. E. 2015. *Organism and Environment: Ecological Development, Niche Construction, and Adaption*. New York, NY: Oxford University Press.

van Ham, R. C. H. J., Kamerbeek J., Palacios C., Rausell C., Abascal F., Bastolla U., Fernández J. M., Jiménez L., Postigo M., Silva F. J., Tamames J., Viguera E., Latorre A., Valencia A., Morán F., Moya A. 2003. Reductive genome evolution in *Buchnera aphidicola*. *Proceedings of the National Academy of Sciences, USA* 100:581–586.

Waddington C. 1942. Canalization of development and the inheritance of acquired characters. *Nature* 150:563–565.

Waddington C. H. 1957. *The Strategy of the Genes*. London: Allen and Unwin.

5
History as Template

5.1 Template

A rather different way of viewing history is as a *template* on which change occurs. Perhaps the most recognizable examples involve the use of a phylogeny—the template—to investigate aspects of the evolutionary history of a group of organisms. An increasingly common approach involves estimating the phylogeny of the group of interest from genetic data, before mapping a second set of data onto this molecular tree. Often the phylogeny itself is of secondary interest; the research is motivated by questions involving other facets of the evolution of the organisms, such as their behavior, their coevolution with parasites or their response to long-distance dispersal.

Importantly, however, these questions can only be answered if we understand the history of the group. Over 30 years ago, as the number of phylogenetic studies based on genetic data was booming, Miles and Dunham (1993: 587) argued that the study of adaptation would greatly benefit from the incorporation of historical, phylogenetic considerations: "an historical approach not only refines the definition and recognition of adaptations, but also provides an objective definition of 'phylogenetic constraints'." The necessity of a phylogeny in understanding adaptation is illustrated nicely in the work of Hamilton (2021), in which she showed that four populations of *Rhagada* land snails in Western Australia have independently evolved a convergent phenotype of a small, banded shell. This parallel evolution has a very different history (and explanation) than the same adaptation in a single clade.

Moreover, we can use our historical approach to explain particular evolutionary changes: unique events need not be beyond the reach of general explanatory methods. Instead of dismissing the derivation of a particular feature as irreducibly contingent (or worse, telling an ad hoc just-so story about it), we can discern its origin and how it later evolved. Let us examine some research exemplifying this approach.

5.2 Evolution of Agonistic Behavior in Birds

Some 60 years ago, the late Gerry van Tets published a monograph about the social displays of a group of birds then classified as the Pelecaniformes: the pelicans, frigatebirds, tropicbirds and the "core Pelecaniformes" (today's Suliformes) the cormorants,

Beyond Equilibria. Hamish G. Spencer, Oxford University Press. © Hamish G. Spencer (2025).
DOI: 10.1093/oso/9780192858993.003.0005

shags, gannets, boobies and darters. This work (van Tets, 1965) was notable for being one of the first to examine the evolution of behavior in an explicitly phylogenetic context: van Tets was interested not only in documenting the highly ritualized communication patterns of the species he studied, but also their evolutionary origins. To do so, he constructed informal phylogenies, hypothesizing how various agonistic displays may have evolved. For instance, van Tets (1965) suggested that the cormorant gaping display, a formalized recognition display between the male and female birds sharing a nest, was derived from fighting and threat behavior, and was related to the head-throwing observed in the red-footed booby, *Sula sula* (see Fig. 5.1).

Subsequently, Kennedy et al. (1996) used parsimony analysis on 37 of these behavioral characters to construct a formal phylogeny, and found that these trees fitted the data well, with statistically significantly structure. Moreover, these behavioral trees were more congruent with independently derived phylogenies based on genetic and osteological data. Hence, contrary to suggestions that behavioral characters are systematically inferior to, say, morphological traits (see arguments refuted in de Queiroz & Wimberger, 1993), Kennedy et al. (1996) were able to show that, in fact, behavior can be phylogenetically informative. And by mapping the behaviors onto a combined genetic and morphological tree, they found that 34 of van Tets's 37 characters were homologous, a rate equivalent to morphological characters.

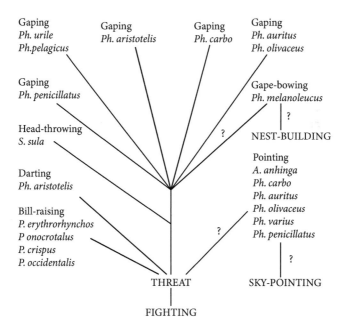

Fig. 5.1 The evolutionary relationships of pelecaniform threat behaviors and their evolutionary derivatives, according to van Tets (1965).

Redrawn from van Tets (1965), with permission.

Kennedy et al. (1996) were also able to make a prediction about the pied cormorant (*Phalacrocorax varius*), a species for which van Tets had not been able to ascertain the presence or absence of gaping. The most parsimonious explanation for the evolution of gaping is that it evolved in the common ancestor of the cormorants and, hence, would be expected to be present in the pied cormorant (unless it had subsequently been lost). Indeed, work subsequent to van Tets's study (summarized in Marchant & Higgins, 1990) confirmed that the pied cormorant does gape. But this analysis can go further, naturally leading to testable predictions about the details of gaping (e.g., the position of the head, whether gaping is repetitive and the form of vocalization during gaping).

Finally, in their discussion of the various forms of wing-waving, Kennedy et al. (1996) noted that the series of evolutionary transformations postulated by van Tets (1965) was more parsimonious than would be expected by chance, although there were, in fact, two alternatives that required slightly fewer changes when mapped onto the combined (genetic + osteological) tree. They examined why van Tets's hypothesis was less parsimonious and found that it was due to the position in the tree of Brandt's cormorant. If this species were sister to the pelagic and red-faced cormorants, with which it uniquely shares "rapid flutter wing-waving," then they pointed out that van Tets's suggestion would be most parsimonious. Subsequent analysis of DNA sequences has shown that, indeed, Brandt's cormorant was misplaced in the combined tree (Kennedy & Spencer, 2014). These three species are each other's closest relatives and, in the latest classifications (e.g., del Hoyo, 2020; Gill et al., 2021; BirdLife International (2021); Harrison et al., 2021), they constitute the genus *Urile*.

In brief, this study shows the power of an explicitly historical approach to understanding the evolution of behavior. At the most general level, it provided robust empirical evidence that behavior can be phylogenetically informative, with some displays persisting for tens of millions of years or more. More specifically, it was able to test behavioral hypotheses about how particular agonistic displays may have evolved in the suliforms, as well as make predictions about missing data. And, in a kind of reciprocal illumination, it raised prescient questions about the accuracy of the template itself.

5.3 Coevolution of Pigeons and Lice

The degree to which hosts and their parasites have coevolved has been hotly debated for well over 100 years. Some parasites exhibit strong host specificity, driven by such factors as host ecology and host-parasite physiological compatibility. As a consequence, their phylogenetic relationships may closely mirror those of their hosts, thus providing evidence of co-speciation (Page, 1994). Although host and parasite phylogenies are not likely to be completely congruent (Fahrenholz's Rule; Fahrenholz, 1913), we can often use various computational tools (reviewed in Filipiak et al., 2016) to quantify the degree of any mismatch as well as to infer possible underlying evolutionary causes (e.g., Page, 2003). For instance, host-switching likely gives rise to different patterns of phylogenetic non-congruence than the extinction of a parasite on a host; both obscure or nullify the imprint of co-speciation (Paterson et al., 2003).

In an intriguing study of co-phylogeny, Clayton and Johnson (2003) compared the evolution of two groups of feather lice that parasitize columbiform birds, pigeons and doves (Fig. 5.2). Although ecologically similar, wing and body lice are morphologically distinct (long and slender versus more rounded, respectively), reflecting the different parts of the bird on which they live and their different strategies to avoid being removed by preening. Importantly, they are phylogenetically separate, monophyletic clades, which means they were able to be treated almost like replicates by Clayton and Johnson. The null expectation under the assumption that the two groups of lice respond in similar ways to speciation in the birds is that the two parasite phylogenies should closely match each other, as well as the tree of the hosts. Nevertheless, subtle ecological differences between wing and body lice as well as chance events might render this prediction void.

Clayton and Johnson (2003) showed that, in some instances but not others, there had been co-speciation between birds and one or other of the louse groups, but that the two lice trees were uncorrelated. They concluded that the factors driving speciation differed between the two groups of lice. More generally, they argued that a failure of lice to speciate when their hosts did was an underappreciated source of host-parasite tree mismatch. (For example, in Fig. 5.2 we can see that *Columbina inca* and *C. passerina*, which are sisters in the bird phylogeny, share a single wing louse species, *Columbicola passerinae*.) They also found that wing lice have parasitized pigeons and doves for a much longer time than these birds' body lice, indeed from an earlier date than the most recent common ancestor of the living columbiforms.

Some of these inferences are, by necessity, specific to this particular bird-louse system and pertain to unique events. Such an example obviously refutes the claims sometimes made that the study of evolution cannot deal with unique events and hence cannot be scientific (see Smith, 2016, for a discussion of such arguments).

Nevertheless, there is clearly a general method being used here (and in similar studies). Finally, the study had implications for methodological research, pointing out the need for tools to detect "failure to speciate."

This sort of co-phylogenetic perspective can be extended to three (and potentially more) levels, and allows insight into questions that at first glance do not have an historical dimension. Grossi et al. (2024), for example, studied the coevolution of several songbird species from South China, their lice from the genus *Guimaraesiella* and the latter's gammaproteobacterial endosymbionts. At least one *Guimaraesiella* species occurs on multiple bird hosts, presumably due to its ability to hitch a ride from bird to bird on hippoboscid flies, which also parasitize birds. As is often the case, the intracellular microorganisms are maternally transmitted, but occasionally they are replaced in the louse cytoplasm by free-living strains. Grossi et al. (2024) used their co-phylogenetic analysis to elucidate, among other things, the likely source, mechanisms and frequency of these replacement events.

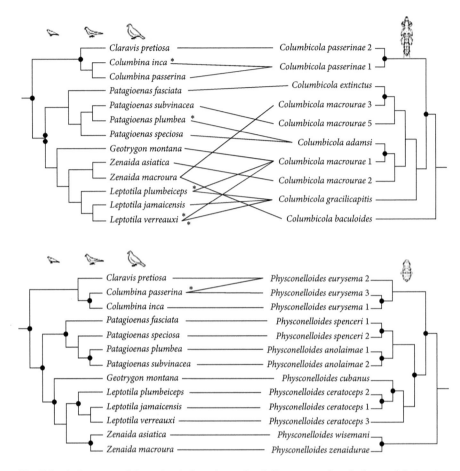

Fig. 5.2 Phylogenies of the columbiform hosts (on left) compared with those of their wing (upper) and body (lower) lice (on right). Host-parasite associations are shown by the lines between the phylogenies.

From Clayton and Johnson (2003), with permission.

5.4 Biogeography of Limpets in the Southern Ocean

Circulation in the Southern Ocean is dominated by the Antarctic Circumpolar Current (ACC) or West Wind Drift, the world's most powerful ocean current, which flows west to east around Antarctica (Fig. 5.3). The ACC has a profound effect on the biota of the Antarctic and Subantarctic regions (Halanych & Mahon, 2018). Recent molecular studies have greatly clarified many aspects of this biota, from species delineation and relationships, to biogeographical patterns and levels of endemism. Here I discuss a pair of studies that reached apparently disparate conclusions about two groups of gastropods. Both studies use novel molecular data to identify and discriminate

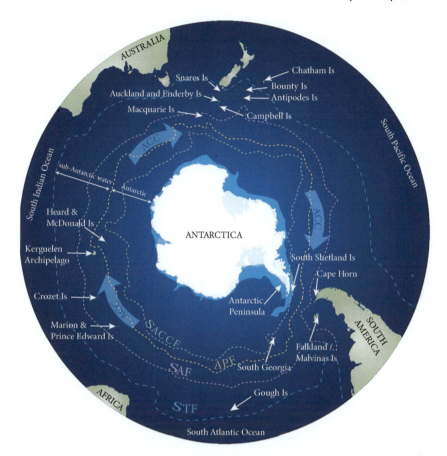

Fig. 5.3 Map of the Southern Ocean, showing the Antarctic Circumpolar Current (ACC), as well as several of the Subantarctic islands mentioned in the text. The Southern Ocean is generally considered to be the seas south of the Subtropical Front (STF). Antarctic and Subantarctic waters meet at the Antarctic Polar Front (APF). Other oceanographic features shown are the Subantarctic Front (SAF) and the Southern ACC Front (SACCF), which delimit the ACC.

Courtesy of C. I. Fraser, with permission.

cryptic species, as well as elucidate their various geographical ranges. The differences in these distributions as well as the corresponding estimates of the respective phylogenies illustrate contrasting histories that have been differentially impacted by ecological factors.

The limpet genus *Nacella* is confined to the cold waters of the Southern Ocean and adjacent seas (González-Wevar et al., 2019). Before the advent of molecular techniques, the taxonomic status and distribution of many of the species had been confused. The 12 species now recognized fall evenly into two clades, one restricted to the coasts of Patagonia (southern Chile and Argentina), the other widely spread around the Southern Ocean, from the Antarctic Peninsula to a number of Subantarctic islands. Each species in this latter clade is narrowly distributed, most being found on a single or a neighboring pair of islands (González-Wevar et al., 2019).

The molecular phylogeny, together with the estimated dates of divergence, show that much of the diversity in the Antarctic/Subantarctic clade is surprisingly recent (Fig. 5.4; González-Wevar et al., 2017). For example, the subclade of *N. edgari* (which lives only on Kerguelen and nearby Heard Islands), *N. macquariensis* (restricted, not unexpectedly, to Macquarie Island) and *N. terroris* (endemic to Campbell Island) shared a common ancestor just ~600,000 years ago, and the last two species only ~250,000 years ago. Similarly, *N. delesserti*, which occurs only on Marion and Crozet

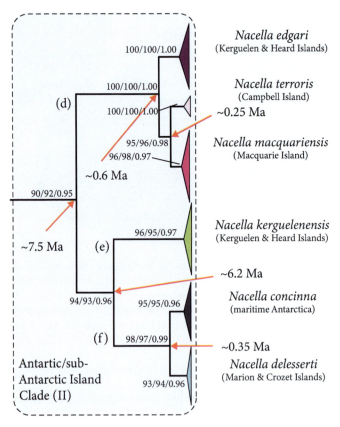

Fig. 5.4 The phylogeny of the Antarctic/Subantarctic clade of the marine limpet genus *Nacella*. The red arrows indicate the estimated dates of divergence of the respective subclades. Note that each species is restricted to a single location, usually one island or island group, with closer locations having more closely related species. These patterns suggest that gene flow is very limited.

After González-Wevar et al. (2017, 2019), with permission.

Islands, is sister to the Antarctic-Peninsula endemic *N. concinna*, and they separated just ~350,000 years ago.

This pattern of recently derived, single-island-group endemics cannot be explained by vicariant processes. Rather, these relationships and their timings are consistent with sequential colonization via rafting in the ACC and subsequent *in-situ* speciation: Kerguelen and Heard Islands, which possess the most distinct of the three species in the first subclade, are upstream of Macquarie Island, which in turn is upstream of Campbell Island; Marion and Crozet Islands are downstream from the Antarctic Peninsula. The rafting vector is highly likely to be the brown macroalga, *Durvillaea antarctica*. Holdfasts of this extremely buoyant and durable bull-kelp harbor numerous invertebrates, including, on occasion, *Nacella* spp. (Powell, 1973). In short, speciation seems to follow hot on the heels of colonization of new islands, from which previously established species have been exterminated, probably by glaciation. Critically, the level of gene flow to these newly colonized islands is not sufficient to prevent speciation, which suggests that successful rafting events are rather infrequent.

The biogeographical patterns in the Southern-Ocean representatives of the pulmonate limpet genus *Siphonaria* are suggestively different. Rather than *Nacella*-like localized endemics, *Siphonaria* comprises two widespread species in the sector from South America to Macquarie Island. Indeed, *S. lateralis* and *S. fuegiensis* are often syntopic, living cheek-by-jowl in the rocky intertidal, especially on and in the holdfasts and stipes of *Durvillaea antarctica* (Fig. 5.5).

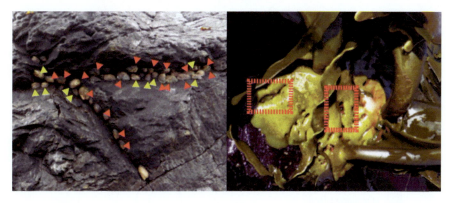

Fig. 5.5 *Siphonaria lateralis* (▲) and *S. fuegiensis* (▲) on rocks (left) and living closely associated with the bull-kelp *Durvillaea antarctica*, Kerguelen Island (right).

From González-Wevar et al. (2018), with permission.

Divergence-time estimates based on sequences of the mitochondrial gene cytochrome-c-oxidase I (COI) suggested that two species separated during the Pliocene, ~4 mya (González-Wevar et al., 2018). There is some degree of genetic differentiation among the populations on different islands (see Fig. 5.6; González-Wevar et al., 2018), but it is not at the level of species. Moreover, populations from some fairly distant places (e.g., Pacific Patagonia and Kerguelen Island) share haplotypes and do so in both species.

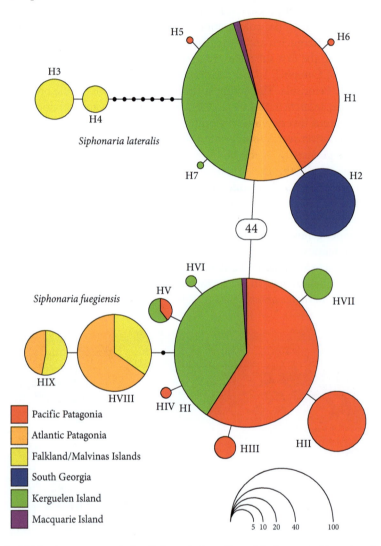

Fig. 5.6 COI haplotype network for *Siphonaria lateralis* and *S. fuegiensis*. Note that the two species are both widely distributed around the Southern Ocean and that distant locations share haplotypes. These patterns suggest frequent, ongoing gene flow.

From González-Wevar et al. (2018), with permission.

In spite of their largely overlapping distributions and habitats, *Nacella* and *Siphonaria* plainly exhibit different levels of diversity. Why? Paradoxically, *Nacella* would appear to have a greater dispersal potential and, hence, population connectivity. *N. concinna* (and presumably its congeners) possess larvae that persist in the plankton for 40–45 days (Peck et al., 2016), and it shows significantly less population structure than the sympatric trochid gastropod *Margarella antarctica*, which possesses benthic-protected development (Hoffman et al., 2011a). Nevertheless, Hoffman et al. (2011b) revealed genetic differentiation between *N. concinna* populations from the Antarctic Peninsula and the South Orkney Islands, a distance of ~600 km, which they argued reflected significant oceanographic barriers to gene flow. Presumably a 45-day planktonic larval stage is insufficient for crossing thousands of kilometers of open ocean.

Siphonaria lateralis, by contrast with *Nacella*, is a direct developer, in which juveniles crawl out from benthic egg masses (Zabala et al., 2020). On first consideration, this aspect of its biology would be expected to impede genetic differentiation, giving rise to a biogeographical pattern more akin to that observed in *M. antarctica*. But, as is depicted in Fig. 5.5, these limpets often live in vast numbers in intimate contact with *Durvillaea antarctica*, which has been well documented as a rafter par excellence. Not only are populations of the bull-kelp itself well connected around the Southern Ocean (Fraser et al., 2009), but so are the populations of many invertebrates with strong associations with the algal holdfasts (Nikula et al., 2010, 2013). It is true that *Nacella* spp. are sometimes found in *Durvillaea* holdfasts, but this relationship is much weaker (Smith & Simpson, 1995). Only the Kerguelen-Heard endemics *N. kerguelenensis* and *N. edgari* (as well as some Patagonian-clade species, such as *N. mytilina*) occur regularly on *Durvillaea* and even then, considerably larger numbers of these limpets live directly on intertidal and subtidal rocks (González-Wevar et al., 2019). As a consequence, gene flow via rafting in *Nacella* is much less than in *Siphonaria*.

In short, history as a template (the *Nacella* phylogeny and the *Siphonaria* haplotype networks) permits us a fuller understanding of the contrasting evolutionary history of these two gastropod genera. Although not precluding some of the inferences we have made, the absence of an historical perspective would mean that we would fail to explain some critical features, such as the close relationship between the species on Macquarie and Campbell Islands, *N. macquariensis* and *N. terroris*, and this pair being sister to *N. edgari* from Kerguelen and Heard Islands.

5.5 Template in Perspective

In providing illuminating explanations for the three exemplars above, ensuring that the temporal dimension was included was crucial. Without it the explanations would have been much weaker or even erroneous. I concur fully with Miles and Dunham (1993: 588) who wrote about studies of the adaptive significance of phenotypic traits, "The absence of an explicit historical perspective weakens the conclusions from comparative analyses."

Looking at the last of my examples above, we can see that one simple ad hoc (and ahistorical) hypothesis for the difference in limpet biogeography is that the *Nacella* radiation was older than that of *Siphonaria*. The use of template in ruling out this explanation focused our attention on the ecology of the limpets and how any differences can impact gene flow and, hence, the evolutionary history of these gastropods.

In addition, the approach used in these studies had a number of wider benefits, such as generating ideas for methodological improvements (in the study of host-parasite co-speciation) or raising questions about the historical template itself (in the cormorant behavioral work). Pleasingly, employing a phylogenetic framework in the analysis of genetic (and other) data is becoming quite standard today, and the above examples are by no means unique. Today, for example, comparative studies routinely factor out the effects of phylogeny.

References

BirdLife International. 2021. IUCN Red List for birds. http://www.birdlife.org [HBW and BirdLife Taxonomic Checklist v5, accessed 22/07/2021].

Clayton D. H., Johnson K. P. 2003. Linking coevolutionary history to ecological process: Doves and lice. *Evolution* 57:2335–2341.

Del Hoyo J. (Ed.) 2020. *All the Birds of the World*. Barcelona: Lynx Edicions.

Fahrenholz von H. 1913. Ectoparasiten und Abstammungslehre. *Zoologischer Anzeiger* 41:371–374.

Filipiak A., Zajac K., Kübler D., Kramarz P. 2016. Coevolution of host-parasite associations and methods for studying their cophylogeny. *Invertebrate Survival Journal* 13:56–65.

Fraser C. I., Nikula R., Spencer H. G., Waters J. M. 2009. Kelp genes reveal effects of subantarctic sea ice during the Last Glacial Maximum. *Proceedings of the National Academy of Sciences, USA* 106:3249–3253.

Gill F., Donsker D., Rasmussen P., (Eds) 2021. *IOC World Bird List* (v11.2). doi: 10.14344/IOC.ML.11.2.

González-Wevar C. A., Hüne M., Rosenfeld S., Nakano T., Saucède T., Spencer H. G., Poulin E. 2019. Systematic revision of *Nacella* (Patellogastropoda: Nacellidae) based on a complete phylogeny of the genus, with the description of a new species from the southern tip of South America. *Zoological Journal of the Linnean Society* 186:303–336.

González-Wevar C. A., Hüne M., Segovia N. I., Nakano T., Spencer H. G., Chown S. L., Saucède T., Johnstone G., Mansilla A., Poulin E. 2017. Following the Antarctic Circumpolar Current: Patterns and processes in the biogeography of *Nacella* (Mollusca: Patellogastropoda) across the Southern Ocean. *Journal of Biogeography* 44:861–874.

González-Wevar C. A., Segovia N. I., Rosenfeld S., Ojeda J., Hüne M., Naretto J., Saucède T., Brickle P., Morley S., Féral J.-P., Spencer H. G., Poulin E. 2018. Unexpected absence of island endemics: Long-distance dispersal in higher latitude sub-Antarctic *Siphonaria* (Gastropoda: Euthyneura) species. *Journal of Biogeography* 45:874–884.

Grossi A. A., Tian C., Ren M., Zou F., Gustafsson D. R. 2024. Co-phylogeny of a hyper-symbiotic system: Endosymbiotic bacteria (Gammaproteobacteria), chewing lice (Insecta: Phthiraptera) and birds (Passeriformes). *Molecular Phylogenetics and Evolution* 190:107957.

Halanych K. M., Mahon A. R. 2018. Challenging dogma concerning biogeographic patterns of Antarctica and the Southern Ocean. *Annual Review of Ecology, Evolution, and Systematics* 49:355–378.

Hamilton Z. R. 2021. Repeated evolution of an undescribed morphotype of *Rhagada* (Gastropoda: Camaenidae) from the inland Pilbara, Western Australia. *Invertebrate Systematics* 35:203–215.

Harrison P., Perrow M. R., Larsson H. 2021. *Seabirds. The New Identification Guide*. Barcelona: Lynx Edicions.

Hoffman J. I., Clarke A., Linse K., Peck L. S. 2011a. Effects of brooding and broadcasting reproductive modes on the population genetic structure of two Antarctic gastropod molluscs. *Marine Biology* 158:287–296.

Hoffman J. I., Peck L. S., Linse K., Clarke A. 2011b. Strong population genetic structure in a broadcast-spawning Antarctic marine invertebrate. *Journal of Heredity* 102:55–66.

Johnson K. P., Clayton D. H. 2003. Coevolutionary history of ecological replicates: Comparing phylogenies of wing and body lice to columbiform hosts. Pages 262–286 in *Tangled Trees: Phylogeny, Cospeciation, and Coevolution*, edited by R. D. M. Page. Chicago: University of Chicago Press.

Kennedy M., Spencer H. G. 2014. Classification of the cormorants of the world. *Molecular Phylogenetics and Evolution* 79:249–257.

Kennedy M., Spencer H. G., Gray R. D. 1996. Hop, step and gape: Do social displays in the Pelecaniformes reflect phylogeny? *Animal Behaviour* 51:273–291.

Marchant S., Higgins P. J. 1990. *Handbook of Australian, New Zealand and Antarctic Birds. Vol. 1. Ratites to Ducks*. Oxford, England: Oxford University Press.

Miles D. B., Dunham A. E. 1993. Historical perspectives in ecology and evolutionary biology: The use of phylogenetic comparative analyses. *Annual Review of Ecology and Systematics* 24:587–619.

Nikula R., Fraser C. I., Spencer H. G., Waters J. M. 2010. Circumpolar dispersal by rafting in two subantarctic kelp-dwelling crustaceans. *Marine Ecology Progress Series* 405:221–230.

Nikula R., Spencer H. G., Waters J. M. 2013. Passive rafting is a powerful driver of transoceanic gene-flow. *Biology Letters* 9:20120821.

Page R. D. M. 1994. Maps between trees and cladistic-analysis of historical associations among genes, organisms, and areas. *Systematic Biology* 43:58–77.

Page R. D. M. (Ed.) 2003. *Tangled Trees: Phylogeny, Cospeciation, and Coevolution*. Chicago: University of Chicago Press.

Paterson A. M., Palma R. L., Gray R. D. 2003. Drowning on arrival, missing the boat and *x*-events: How likely are sorting events? Pages 287–309 in *Tangled Trees: Phylogeny, Cospeciation, and Coevolution*, edited by R. D. M. Page. Chicago: University of Chicago Press.

Peck L. S., Heiser S., Clark M. S. 2016. Very slow embryonic and larval development in the Antarctic limpet *Nacella polaris*. *Polar Biology* 39:2273–2280.

Powell A. W. B. 1973. *The Patellid Limpets of the World (Patellidae). Indo-Pacific Mollusca 3*. Auckland: Auckland Institute and Museum.

de Queiroz A., Wimberger P. H. 1993. The usefulness of behaviour for phylogeny estimation: Levels of homoplasy in behavioural and morphological characters. *Evolution* 47:46–60.

Smith R. J. 2016. Explanations for adaptations, just-so stories, and limitations on evidence in evolutionary biology. *Evolutionary Anthropology* 25:276–287.

Smith S. D. A., Simpson, R. D. 1995. Effects of the 'Nella Dan' oil spill on the fauna of *Durvillaea antarctica* holdfasts. *Marine Ecology Progress Series* 121:73–89.

van Tets G. F. 1965. A comparative study of some social communication patterns in the Pelecaniformes. *Ornithological Monographs* 2:1–88.

Zabala S., Averbuj A., Bigatti G., Penchaszadeh P. E. 2020. Embryonic development of the false limpet *Siphonaria lateralis* from Atlantic Patagonia. *Invertebrate Biology* 139:e12276.

6

Chance, Chaos and Capriciousness

6.1 Chance, Chaos and Capriciousness

Three further flavors of history—*chance, chaos* and *capriciousness*—are often manifested in superficially similar ways. Nevertheless, they are conceptually distinct. Chance is unique in being random; it is about historical events that cannot be predicted. Chaos can certainly look random and may, in practice, be difficult to distinguish from chance. And, paradoxically, capriciousness violates the "usual" rules of random behavior and is decidedly not random.

Chance goes hand-in-hand with unpredictability, as in the paradigmatic case of radioactive decay. We cannot tell exactly when the nucleus of an atom of uranium-232, for example, will emit an alpha particle. Some events may seem to be random because they are particularly complex, with so many causal factors and constraining conditions, but I suggest these do not fit under the flavor of chance, since investigation of the details may show other flavors, most notably contingency and constraint, at work. Nevertheless, whether chance or some other flavor is the best descriptor of the history in a particular system is, I think, a secondary issue; what really matters is the historical perspective.

Chance is invoked, too, in much of statistics, where the "error term" explains the difference between the prediction of the statistical model and the observed value. This term is conceptually last in the explanation, with the model's main factors (and maybe their interactions) doing the heavy lifting. It's as if some dice-throwing determines the final outcome after the real biology has played out. Again, this kind of chance is not really chance as history; rather it arises because we do not have a full explanation of our biological system.

In ecology and evolution, the best examples of history as chance are probably genetic drift and mutation. Genetic drift arises from the stochastic sampling of alleles during reproduction, and mutation, too, appears random: we cannot predict if a particular allele will mutate to some other version of that gene, although we may be able to modify the mutation rate.

Chaos, however, is completely deterministic. If you re-run a chaotic model with exactly the same input, you will get identical output every time; a stochastic model will always (well, almost always) give a different result. Chaotic systems are extremely sensitive to changes in the values of variables and parameters: small differences lead

Beyond Equilibria. Hamish G. Spencer, Oxford University Press. © Hamish G. Spencer (2025).
DOI: 10.1093/oso/9780192858993.003.0006

to very different dynamics and/or equilibria. It is this sensitivity that leads to the similarity with chance, since it makes the chaotic system appear unpredictable, especially in the longer term. Biological examples of chaos are often complex: the best understood are models of the dynamics of disease outbreaks, which depend on a myriad of parameters and often contain numerous variables.

For example, the model of Eilersen et al. (2020), which examines one predator and two prey species, of which one of the latter is susceptible to disease, is relatively simple. It is an extension of the well-known Lotka-Volterra system to include two prey species, combined with the standard SIR (Susceptible, Infectious, Recovered) model of disease. Yet it has nine parameters and four variables, even though none of the three species has a carrying capacity and there is no competition between prey species. Conceptually, therefore, the model, which manifested chaos over a wide range of parameters, could be far more complicated. Nevertheless, chaos can arise in remarkably simple models (see below); chaos does not require biological complexity.

Lewontin (1966) first defined capricious in comparison with deterministic and stochastic. A deterministic model or system is (as we saw in Chapter 2) one in which we know precisely what is happening; we have perfect information and can predict exactly what will happen next. Stochastic models have that element of the unknown because of chance. Nevertheless, as the number of observations of the system increases (or as the model produces more outcomes), the more we know and, indeed, we approach perfection. A capricious system is one in which no matter how many samples we have, our knowledge of the system fails to increase. To paraphrase Lewontin, capriciousness is less certain than chance.

Lewontin (1966, 1967) illustrated capriciousness with two computer simulations of a simple population-genetic model with the same initial allele frequency and the same set of varying selection pressures. These two runs differed only in the order of the environmental states, which affected the selection coefficient in the model, and yet they exhibited very different dynamics, ending up at different allele frequencies. This contrast implied, amongst other things, that there was no average selection coefficient. I recapitulate this example in section 6.4.

6.2 Chance

In common parlance, we often ascribe the cause of unexpected events to chance. Maybe we are a little sophisticated in describing the particular occurrence as an outlier, implying some probability distribution of potential outcomes. It seems to me, however, that we often invoke chance as a "cause of last effect," when our other explanations fail. The approach to statistical model-fitting described above is a prime example here.

A canonical example is the model we have for quantitative-trait variation in natural populations, in which three factors (and their interactions) are held to be causal. Two of these factors—genetic variation among individuals and environmental heterogeneity—are uncontroversial, but the third—developmental noise—is likely a grab-bag of different processes, called upon to explain differences that we cannot attribute to genes, environment or their interaction. Some of these noisy processes are probably truly random, involving the stochastic nature of biochemical reactions, but others may simply be poorly understood aspects of development. For example, the effect of the microbiome, which often has a degree of inheritance and can interact with genes and environment, has only been seriously examined more recently (see Berg et al., 2020, and references therein). These effects, which are neither genetic (in spite of their inheritance) nor environmental (because of their inheritance) would be ascribed to noise (or chance), whereas they are not actually chance, but due to a causal factor that is, to some degree at least, predictable.

Nevertheless, chance is unequivocally an important historical flavor. And, as I indicate above, even if chance is sometimes a cloak for our ignorance of the details of history, it is better to recognize the historical aspect of chance than ignore it.

The theory of population genetics probably encompasses the most in-depth treatment of chance in ecology and evolution. Genetic drift has at least three potential underlying causes: (i) the chance events that determine which alleles end up in which meiotic products (especially when not all these products become gametes), (ii) the random nature of which individuals contribute to the pool of gametes and how much they do so, and (iii) the random sampling of gametes that unite at fertilization. Nevertheless, each one of these aspects of drift may be influenced by non-random forces. For example, fertilization may not occur at random, with complex interactions between male and female molecules (often proteins) that mediate sperm competition (Chow et al., 2010). And we have long known about meiotic drivers, genetic elements that "cheat" during meiosis, so that they are over-represented in the viable gametes, sometimes to the virtual exclusion of their rivals (Sandler & Novitski, 1957; Lindholm et al., 2016).

All these caveats aside, recognizing the effects of genetic drift has been fundamental to our understanding of evolution. The classic demonstration of genetic drift by Buri (1956) serves to illustrate some important features of history as chance. In brief, Buri set up 107 replicate populations of *Drosophila*, each with eight males and eight females heterozygous for an eye-color allele, bw^{75} (and hence an initial allele frequency of 0.5). To initiate the next generation of each replicate, he randomly sampled eight males and eight females from the offspring of that replicate and recorded their genotypes (heterozygotes were distinguishable from both homozygotes). Thus, each generation of each replicate consisted of 32 genes; Buri (1956) reported the numbers of the bw^{75} allele for the following 19 generations. Results from selected generations are shown in Fig. 6.1.

The allele frequencies in the different replicates immediately spread out from the initial 0.5 and by generation 19, more than half had either lost or fixed the bw^{75} allele. The random sampling of flies (and hence the different alleles) each generation produces a random walk in the allele frequency that, in the long term, inevitably results in the fixation or loss of the allele. That chance is involved can be seen in the different

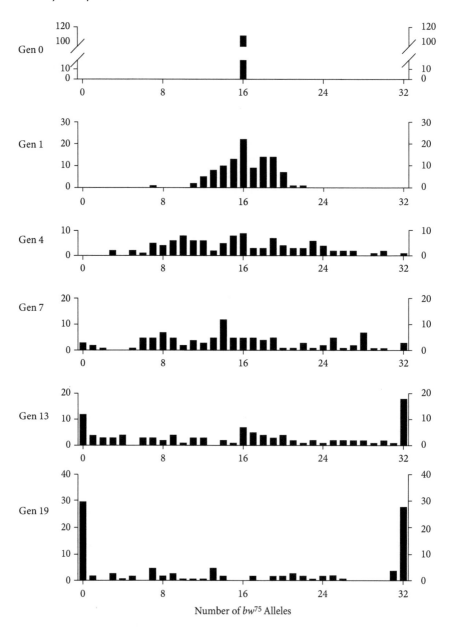

Fig. 6.1 The number of replicate *Drosophila melanogaster* populations (out of 107) with different numbers of bw^{75} alleles at selected generations in the experiment of Buri (1956). In the initial Generation 0, all populations comprised 16 heterozygotes and so, of the 32 genes present, 16 were bw^{75} alleles. Each generation consisted of a random sample of eight males and eight females, modeling genetic drift. By Generation 19, almost 30 populations had lost all bw^{75} alleles and a similar number fixed for this allele. Some populations remained polymorphic, however.

outcomes in the different replicates: the allele frequencies do not change in a deterministic manner. Moreover, given the frequency in one generation, we cannot tell (unless fixation or extinction has already occurred) whether the following generation will exhibit a larger, smaller or the same frequency. The direction of change is not contingent on the current frequency, nor is it constrained by it. Chance is not contingency or constraint or template.

Nevertheless, the process is sufficiently well behaved that we can make predictions about the ensemble of replicates, as well as individual replicates in the long term. For example, each replicate will eventually fix or lose the bw^{75} allele; given the initial frequency of 0.5 and the absence of selection, half the replicates are expected to lose the allele and half fix it; we can predict how many generations this loss of variation will take, on average. We can even construct a distribution of frequencies likely in the next generation, given the frequency in the current generation. In short, chance events are amenable to statistical analyses, which affords scientific investigation and some degree of explanation.

The chief difficulty with statistical explanations of a process such as genetic drift is that we are often interested in just a single instance or observation. Suppose we knew of just one of Buri's replicates, which happened to have lost the bw^{75} allele in generation 13. What can we conclude about the process? In short, we can say rather little: we can certainly see, given the fluctuations in allele frequency from generation to generation, that genetic drift has had an effect. But we cannot be sure that the loss is due solely to drift, since we cannot rule out the alternative (or possibly additional) explanation that selection was acting against the allele. And if we had knowledge of two replicates, one that fixed and one that lost bw^{75}, we could not conclude (much as we might naively be tempted) that the difference was a consequence of small habitat differences that favored alternative alleles. Chance inherently leads to different possible outcomes, which is one reason it is good to be sure it really is that flavor of history acting in the study system.

Chance can be superimposed on an underlying deterministic process. (In an odd way, Buri's experiment can be seen as drift imposed over selection, in which selection was absent. But Buri did a second set of experiments in which it was clear that the bw^{75} allele was selectively favored even as drift was also present: the allele frequencies changed as a consequence of two causes.) Moreover, chance will often interact with these non-random processes to produce a unique outcome.

As an example, consider the simple population-genetic model of selection against a deleterious recessive allele, a, which we first met in Chapter 2. AA and Aa genotypes produce phenotypes with relative fitness 1, whereas aa genotypes have phenotypes with fitness $1 - s$ (with $0 < s \leq 1$). In an infinite population (i.e., in the absence of genetic drift), the allele frequency of a then iterates according to

$$q' = \frac{q(1 - sq)}{1 - sq^2}.$$

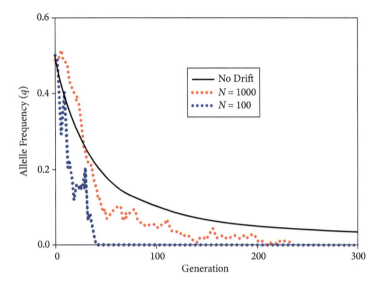

Fig. 6.2 Allele frequency over time in a simple model of genetic drift and selection against a recessive allele. Drift and selection interact, and the allele is lost faster in smaller populations.

As is exemplified by the solid black line in Fig. 6.2, the frequency of *a* rapidly decreases from any initial large value, but as it approaches zero, change slows down dramatically. When *a* is rare, most of these alleles are found in heterozygotes and so are masked from the effects of selection. In a finite population, however, drift becomes very important when *a* is rare, greatly speeding up the eventual extinction of the allele, as is shown by the red and blue dotted lines. Any random decreases due to drift are maintained (or even enhanced) by selection, whereas any increases are opposed (albeit weakly).

Once again, our understanding of this extremely simple system is enhanced by knowledge of its history. We might be tempted to claim that extinction of the *a* allele in all the cases is due to selection, but recognition of the history reveals that, for the red and blue populations, the true explanation involves the interaction of selection with the chance effects associated with genetic drift.

6.3 Chaos

Instead of dismissing history as "chance," we could lazily say it was "chaotic." Both descriptors imply unpredictability, but in spite of this similarity, they are formally fundamentally different. Chaotic systems are deterministic but their long-term behavior appears unpredictable because of their sensitivity to initial conditions:

small changes can have dramatically large consequences after a few iterations of the equation(s) governing the system.[1]

Remarkably, even apparently straightforward systems can be chaotic. The well-known logistic iteration for single-species population growth with discrete generations, for example, is often written

$$N_{t+1} = N_t \left(1 + r\left(1 - \frac{N_t}{K}\right)\right),$$

where r is the intrinsic growth rate of the population and K is its carrying capacity. If we let $x_t = \frac{r}{1+r}\frac{N_t}{K}$ and $a = 1 + r$, we can rewrite our equation as

$$x_{t+1} = a x_t (1 - x_t), \tag{6.1}$$

which is more convenient for mathematical analysis and reveals the extraordinary simplicity of the example. We confine ourselves to values of $x_t \in [0, 1]$ and $a \in [1, 4]$.

As May (1975) showed, different values of a give utterly different, albeit completely determined, behaviors, as shown in Fig. 6.3. (In other words, the model is structurally unstable as the value of a is altered.) For $a = 2.0$, the system quickly iterates to a globally stable equilibrium value of $x_t = 0.5$; for $a = 3.4$ there is a two-cycle, the value of x_t oscillating between 0.452 and 0.842. A further increase in a, for instance to 3.5, gives a four-cycle, with x_t cycling around four different values. For $a = 3.8$, however, there is no equilibrium value, nor a regular cycle; the system is aperiodic, careering around seemingly random (but in fact completely predictable) values. In this last instance, even close-by values of x_t lead to very different iterated values within just a few generations.

So why does all of this matter? To paraphrase May (1975), it is disturbing (May's word) that such a simple and completely predictable, single-species model of population could potentially exhibit arbitrary behavior. Moreover, as May (1975) noted, such bizarre behavior is not limited to the specific example of Equation 6.1; almost any density-dependent model has similar properties (May & Oster, 1976). Such findings imply that for sufficiently large population growth rates (i.e., values of r and hence a), density-dependent regulation could yield dynamics that mimicked chaos. In practice, such a time series might be explained as being due to chance, even if it was actually completely deterministic.

Fortunately, however, the deterministic nature of chaos does mean some predictions can be made, at least in the short term, and this approach allows chaos to be distinguished from chance. Sugihara and May (1990), for instance, outlined a nonparametric method for predicting the next value (or values), given the values already

---

[1] Technically, there also needs to be "topological mixing," which means that iterations started with values from any region of state space will eventually reach any other region. In some cases, such as the example developed below, this mixing may occur only over a subset of state space and so, strictly, the system is chaotic only over that subset.

74 • Beyond Equilibria

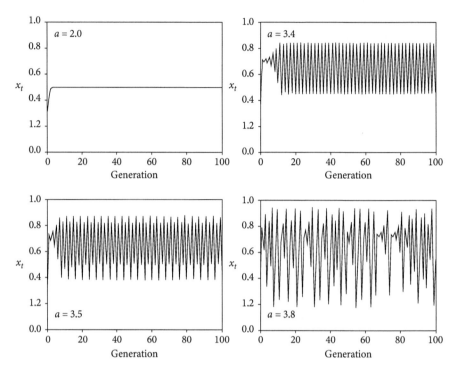

Fig. 6.3 Iterations of Equation 6.1, starting at $x_0 = 0.3$, for increasing values of a. The four plots show a globally stable equilibrium, a two-cycle, a four-cycle and, finally, chaos as a increases. (Technically, there is chaos only for $x_t \in \left[\frac{a^2}{4}\left(1 - \frac{a}{4}\right), \frac{a}{4}\right]$).

known. They showed that prediction further into the future became less accurate, which was not the case with a noisy limit cycle (i.e., uncorrelated random values added to a regular cycle). This difference, they suggested, could be a clue to the presence of deterministic chaos.

Although the proof of principle was illustrated with simulated data, this method was then also applied to real data: measles and chickenpox case numbers in New York City. The measles data had previously been interpreted variously as a noisy limit cycle, as well as chaos on top of a seasonal cycle; Sugihara and May's (1990) analysis added to the case for the latter. By contrast, the chickenpox data was better explained by the former model. In other words, the measles epidemic was governed by a deterministic process, whereas the chickenpox numbers were stochastic overlaying a deterministic process.

Examining history seriously and using the appropriate analytical tools leads to significantly better understanding. This realization means that we cannot ascribe apparently unpredictable fluctuations in a variable over time to chance, whether that is stochasticity directly or randomness in (say) environmental conditions or other parameters. Nor can we avoid the issue completely and claim without further evidence that these observations are due to observational or experimental error. More practically, differences such as those between the measles and chickenpox data imply that the transmission dynamics of these two illnesses are distinct, which presumably has consequences for disease control.

6.4 Capriciousness

The idea that more data does not improve our knowledge of a system seems an anathema to the scientific method! But mathematicians have long known that certain "pathological" examples display such counter-intuitive properties. In statistics, the Cauchy distribution is one such case, even though it is superficially very similar to the normal (Gaussian) distribution (Fig. 6.4), indeed distressingly so for scientists who might want to distinguish them. Moreover, it can arise quite naturally, for example, as the ratio of two independent normally distributed random variables.

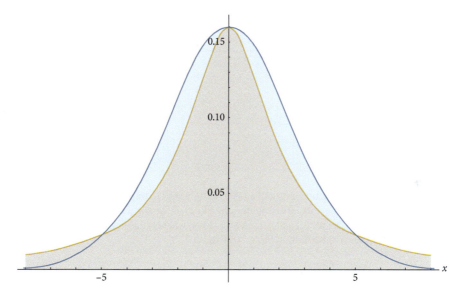

Fig. 6.4 The probability density function of a Cauchy distribution (brown) with location parameter 0 and scale parameter 2 compared to that of a normal distribution (blue) with mean 0 and variance 2π. The Cauchy looks rather like a normal with more density in the tails.

The normal distribution is well known to behave impeccably, whereas the Cauchy distribution, bizarrely, doesn't even have a mean or variance. (Of course, a *sample* from a Cauchy distribution has a *sample* mean and *sample* variance.) Unlike the normal distribution, which is parametrized by its mean and variance, the

Cauchy distribution is determined by a "location parameter" and a "scale parameter" (see Fig. 6.4).

Fig. 6.5 illustrates the capricious nature of the Cauchy distribution, plotting the sample mean of a single sample as it becomes larger and larger. For a well-behaved distribution like the normal, even the means of small samples are close to the true mean, with larger samples' means even closer to that value. But for the Cauchy, larger samples do not contain more information about the location parameter (which naively might be thought to be the mean). In the instance illustrated in Fig. 6.5, even a sample size of 10^5 has a mean clearly greater than two, nowhere close to the location parameter of zero. The means of larger and larger samples simply do not approach any particular value.

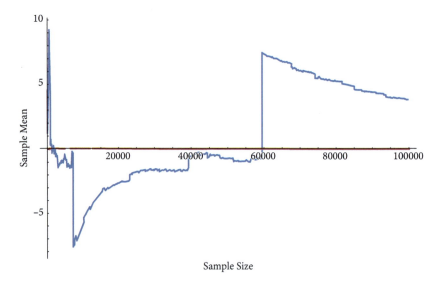

Fig. 6.5 The arithmetic sample mean of an increasingly larger sample from the Cauchy distribution (blue line) and the normal distribution (red line) of Fig. 6.4. The red line starts off close to zero (the normal's true mean) and iterates ever closer, whereas the blue line has no long-term trend, since the Cauchy distribution has no true mean.

So where in biology does capriciousness arise? As mentioned above, Lewontin (1966, 1967) argued that it was commonplace in standard population-genetic models. The details of his example (Lewontin, 1967) appear to have an error, so I have reworked his example. The simple model of selection for or against an allele of additive effect can be parametrized by assigning fitnesses of $1 + s$, 1 and $1 − s$ to the three genotypes AA, Aa and aa, respectively, where $s \in [−1, 1]$. With discrete generations, the frequency of the a allele, q, iterates according to

$$q' = \frac{q(1 - sq)}{1 + s(1 - 2q)}.$$

Supposing the environment varies, so that the value of s changes each generation in a random uncorrelated manner. Sampling independent values from the uniform distribution on the interval from −1 to 1 gives us such a scenario. The red line in Fig. 6.6 shows the frequency of the a allele over 50 generations of selection for a particular series of s values, starting at a frequency of $q = 0.5$.

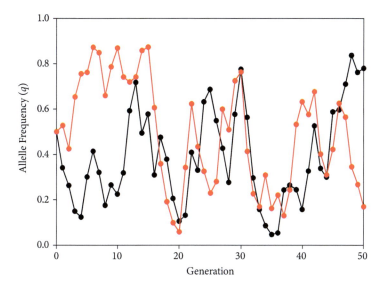

Fig. 6.6 The allele frequency in a simple model of selection in which the selection coefficient (s) changes randomly each generation. The black line shows the allele-frequency trajectory for the same series of s values as the red line, but in the reverse order.

But what if we had the exact same set of s values, but in the reverse order? The black line shows that allele-frequency trajectory over 50 generations. It is obvious that, in spite of experiencing the exact same environments but in a different order, the outcome is utterly different. Lewontin (1966) called this dependence on the order of the environments the "principle of historicity."

The different outcomes mean that there is no "average" selection coefficient that suffices to describe what has happened. Indeed, even though the mean value of s in this particular case is close to zero (in actual fact it was ~0.05), one series of environmental states reduced the frequency of a to 0.17, whereas the exact same states in reverse increased it to 0.78. Remember, this is a deterministic model (in spite of the initial random generation of the sequence of s values); there is no genetic drift.

Lewontin argued that such behavior was capricious, worse even than stochastic. If we had observed the allele frequencies for both runs in generations 0 and 50, we might have tried to say that the outcomes differed because of genetic drift, i.e., randomness. But there is no drift here and no chance differences between the two populations, both of which are governed by a completely deterministic model.

Perhaps more likely, we would have tried to explain what was going on in terms of some form of selection for *a* in the black population; against it in the red. But actually, what matters is really what has happened just recently, say the last five or so generations, where it is true that *a* was favored in the black population and deleterious in the red. This system is capricious because we cannot describe how it has behaved by considering some average action of selection.

Capriciousness describes systems in which there is poor memory: history matters, but ancient history less so. Lewontin (1966) viewed a capricious system as one that forgets; hence new data does not lead to an increase in information. By contrast, systems that are well behaved statistically, obeying the law of large numbers, approach perfect information as the amount of data increase.

Nevertheless, the underlying process is Markovian: the next generation's allele frequency depends only on its current frequency (and the value of *s*), not any of the past values. The reason for capriciousness in this example becomes clearer when we examine the formula for the change in allele frequency:

$$\Delta q = q' - q = \frac{-sq(1-q)}{1 - sq + s(1-q)}.$$

Fig. 6.7 plots this equation as a function of *q*, from which it can readily be seen that values of *q* close to zero or one produce small values of Δq regardless of the size of *s*. Thus, populations closer to loss or fixation of *a* are less affected by selection,

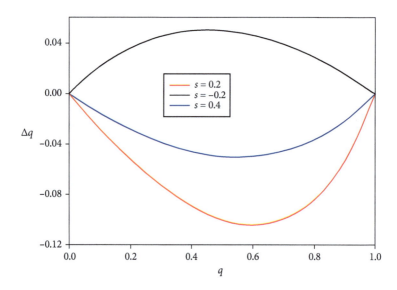

Fig. 6.7 The change in allele frequency in a simple model of selection for different values of the selection coefficient (*s*). Intermediate allele frequencies give larger changes, irrespective of *s*.

experiencing a kind of inertia in their allele-frequency changes that populations with intermediate frequencies do not.

6.5 Chance, Chaos and Capriciousness in Perspective

Chance, chaos and capriciousness form a trio of difficult-to-analyze flavors. Even though they are conceptually separable and imply very different underlying causes of change in our study systems, their practical separation is not always straightforward. What they have in common, however, is the potential to suggest simpler outcomes than is warranted. Differences in allele frequencies among populations may look as if they are due to selection, for instance, whereas in fact they may be due to genetic drift (i.e., chance), small initial differences in the system (i.e., chaos) or simply the order of causative forces (i.e., capriciousness).

Of course, chance, chaos and capriciousness can all contribute when other historical flavors are present (see Chapter 10 for a fuller discussion). In particular, various of these three flavors can all be contributing to the dynamics of a system at any one time. It is not hard to imagine a capricious system of selection on allele frequencies in which chance in the form of genetic drift is also acting. Or perhaps there is a chaotic predator-prey-disease model, on which demographic chance is overlaid.

References

Berg G., Rybakova D., Fischer D., Cernava T., Vergès M. C., Charles T., Chen X., Cocolin L., Eversole K., Corral G. H., Kazou M., Kinkel L., Lange L., Lima N., Loy A., Macklin J. A., Maguin E., Mauchline T., McClure R., Mitter B., Ryan M., Sarand I., Smidt H., Schelkle B., Roume H., Kiran G. S., Selvin J., Souza R. S. C., van Overbeek L., Singh B. K., Wagner M., Walsh A., Sessitsch A., Schloter M. 2020. Microbiome definition re-visited: Old concepts and new challenges. *Microbiome* 8:103.

Buri P. 1956. Gene frequency in small populations of mutant Drosophila. *Evolution* 10:367–402.

Chow C. Y., Wolfner M. F., Clark A. G. 2010. The genetic basis for male × female interactions underlying variation in reproductive phenotypes of Drosophila. *Genetics* 186:1355–1365.

Eilersen A., Jensen M. H., Sneppen K. 2020. Chaos in disease outbreaks among prey. *Scientific Reports* 10:3907.

Lewontin R. C. 1966. Is nature probable or capricious? *BioScience* 16:25–27.

Lewontin R. C. 1967. The principle of historicity in evolution. Pages 81–94 in *Mathematical Challenges to the Neo-Darwinian Interpretation of Evolution*, edited by P. S. Moorhead and M. M. Kaplan. Philadelphia: Wistar Institute Press.

Lindholm A. K., Dyer K. A., Firman R. C., Fishman L., Forstmeier W., Holman L., Johannesson H., Knief U., Kokko H., Larracuente A. M., Manser A., Montchamp-Moreau C., Petrosyan V. G., Pomiankowski A., Presgraves D. C., Safronova L. D., Sutter A., Unckless R. L., Verspoor R. L., Wedell N., Wilkinson G. S., Price T. A. R. 2016. The ecology and evolutionary dynamics of meiotic drive. *Trends in Ecology & Evolution* 31:315–326.

May R. M. 1975. Biological populations obeying difference equations: Stable points, stable cycles, and chaos. *Journal of Theoretical Biology* 51:511–524.

May R. M., Oster G. F. 1976. Bifurcations and dynamic complexity in simple ecological models. *American Naturalist* 110:573–599.

Sandler L., Novitski E. 1957. Meiotic drive as an evolutionary force. *American Naturalist* 9:105–110.

Sugihara G., May R. M. 1990. Nonlinear forecasting as a way of distinguishing chaos from measurement error in time series. *Nature* 344:734–741.

7

Approach and Turnover

7.1 Approach and Turnover

Approach and *turnover* are the names I have given to two flavors of history that explicitly acknowledge that the system may not be at an equilibrium. The flavor of approach is present when the equilibrium is not yet reached, maybe because it is only slowly being approached. As a consequence, the system spends much time—maybe even almost all the time—not at the actual equilibrium. Maybe the equilibrium is like a poor magnet, attracting iron filings only weakly. And, indeed, disturbances to the system—occasional gusts of wind blowing the metal dust away from the magnet—may mean that equilibrium is never reached or, if it is, that the system is easily displaced from equilibrium. As already mentioned, and more fully developed below, the equilibria in the standard population-genetic models of mutation-selection balance provide good examples of approach.

By contrast, turnover occurs when the system has no equilibrium or, perhaps more accurately, when elements of the system are changing but give the impression of stasis. Turnover is rather like a swan paddling furiously below the water in order to maintain its graceful stationary position in the stream. Two enormously important models in ecology and evolution—island biogeography and the neutral theory in population genetics—epitomize turnover. On the surface things are calm: the number of species on the island reaches some steady state, as does the expected level of heterozygosity, respectively. Nevertheless, below the surface everything remains dynamic: the species composition of the community on the island remains in flux, while different neutral mutations arise, drift to extinction or fixation and are, inevitably, later displaced.

7.2 Approach

Almost any system with an equilibrium will take some time to reach that state. True, some systems starting out away from equilibrium move immediately there, but these are the exception, the extremum of a continuous range that extends, eventually, to those systems too far from an equilibrium to feel any attraction. Instead, most systems take time to arrive at equilibrium. Moreover, as we saw with the model of selection against a deleterious recessive allele in an infinite population in Chapter 6, things may slow down as the equilibrium is neared. Two empirical questions consequently arise: How long will it take to arrive at equilibrium? And, how close to equilibrium does the system need to be for the difference to be inconsequential?

7.2.1 Mutation-selection Balance

Let us examine my archetypal case, mutation-selection balance. These models, some of the oldest in population genetics, were derived in order to explain the presence at non-negligible frequencies of alleles with distinctly negative fitness effects, especially those underlying human diseases such as phenylketonuria (PKU). (Such diseases are frequently, if inaccurately, described as "genetic diseases;" see Zuk & Spencer, 2020.) They envisage an equilibrium between selection reducing the frequency of the disease gene (the mutation or mutant allele) and the process of mutation increasing it.

The details of the model and some numerical examples are given in Box 7.1. In short, for a deleterious recessive, the equilibrium allele frequency is given by $q = \sqrt{\mu/s}$, where μ is the mutation rate and s is the strength of selection against homozygotes for the mutant allele. Moreover, once the population reaches a low frequency of the mutation, the changes in frequency become very small. Further decreases due to selection are hindered by the presence of the vast majority of deleterious alleles in selection-immune heterozygous reservoir and any increases due to mutation are small by virtue of the low mutation rate. Any real population, therefore, may well not be at the value predicted by the model.

Box 7.1 Classical Mutation-selection Balance for a Deleterious Recessive Allele

Suppose we have a single di-allelic locus, at which homozygotes for the recessive a allele suffer a fitness deficit of s compared to homozygous AA and heterozygous Aa individuals. As we saw in section 6.2, under a model of discrete generations, selection will reduce the frequency of the a allele from q to q^s according to the iteration

$$q^s = \frac{q(1-sq)}{1-sq^2}.$$

Recurrent mutation from A to a at a rate μ each generation, however, will increase the frequency of a to q':

$$q' = q^s + \mu(1 - q^s).$$

Since we are trying to explain the frequency of a rare allele, it is reasonable to ignore any back-mutation from a to A. The rarity of a means that back-mutation will be of little effect. Intuitively, at some point there should be a balance between the selective reduction in the frequency of a and its mutational increase. Mathematically, we want to find the value of q satisfying $q' = q$. A little algebra then gives three solutions: (i) $q = 1$, corresponding to fixation of a, which is biologically unrealistic; (ii) $q = -\sqrt{\mu/s}$, which is biologically impossible; and (iii) $q = \sqrt{\mu/s}$, which is the well-known formula we seek.

Box 7.1 *Continued*

To determine how quickly this equilibrium value is reached, we first approximate the overall iteration near the equilibrium by a linear equation and then find the leading eigenvalue of the system. See mathematical textbooks such as Edelstein-Keshet (1988) and Otto and Day (2007) for clear and detailed explanations of these techniques. We have, therefore,

$$q' = \frac{q(1 - sq) + \mu(1 - q)}{1 - sq^2}$$

$$\approx \sqrt{\frac{\mu}{s}} + \frac{1 + \mu - 2\sqrt{\mu s}}{1 - \mu}\left(q - \sqrt{\frac{\mu}{s}}\right).$$

If we arrange this linearization to read

$$q' - \sqrt{\frac{\mu}{s}} \approx \frac{1 + \mu - 2\sqrt{\mu s}}{1 - \mu}\left(q - \sqrt{\frac{\mu}{s}}\right),$$

we can see that, near the equilibrium, the distance from it is reduced each generation by a constant factor (the eigenvalue) of $\frac{1+\mu-2\sqrt{\mu s}}{1-\mu} \approx 1 - 2\sqrt{\mu s}$ (since μ is very small), which is very close to one. (For instance, if $s = 0.2$ and $\mu = 5 \times 10^{-6}$, the value is 0.998.)

Accordingly, there is very little reduction at all, and the approach to equilibrium is glacial. (To continue our previous numerical example, the equilibrium value is ~0.005; a population with a frequency of 0.010 would decrease to ~0.00998; the full model gives 0.00998515.)

Nevertheless, it might be argued that the differences are trivial: does it truly matter if the frequency is 0.010 rather than 0.005? The answer depends on the issue being examined. In this numerical example, for instance, an allele frequency of 0.010 would suggest that one in 10,000 (0.010^2) live births would manifest the condition; if the frequency were the equilibrial value of 0.005, the incidence would be just one quarter of that value. And if the mutational frequency of 0.010 was mistakenly thought to be the equilibrium value and used to estimate the mutation rate, this calculation would give $\mu = 2 \times 10^{-5}$, a value four times too large. If these erroneous values were used, say, in genetic counseling, they may well matter.

The alluring nature of ahistorical explanations is revealed by this simple example. Almost all population-genetics textbooks (e.g., Charlesworth & Charlesworth, 2010; Nielsen & Slatkin, 2013) quote (or derive) the $q = \sqrt{\mu/s}$ formula. Rarely, however, do these same authorities note that the rate of approach to this equilibrium value is painfully slow. Readers are thus left in a position to draw the sort of erroneous conclusions outlined in the previous paragraph.

As Kingsland (1985) has argued so eloquently, the focus on equilibria and their mathematical tractability can cloud our view of some interesting and relevant biology. I am by no means the first to make this point, even in a model involving mutation: 100 years ago, Fisher (1924) noted that omitting proper consideration of the rate of approach to equilibrium has significant consequences and it may well be more important than the value of the equilibrium itself.

7.2.2 Masking Tay-Sachs in NYC

A concrete and more recent example involves suggestions for reducing the incidence of infantile Tay-Sachs disease in ultra-Orthodox Jews in New York City. Like PKU, Tay-Sachs is often described as a genetic disorder, since it occurs only in homozygotes for a rare allele, heterozygous carriers being unaffected[1]. Infantile Tay-Sachs has a devastating impact, becoming apparent in the first year of life, with progressive neurological deterioration leading to cerebral seizures, severe mental impairment, loss of vision and motor control, and eventually death, usually by age six. Although rare in most parts of the world, the disease is much more frequent in certain populations and is of real concern. In Ashkenazi Jews from New York City, q, the frequency of the defective allele, is estimated at 1/16, meaning a little under half of one percent of babies—one quarter of the offspring of two heterozygous parents—would be expected to have the disease.

Merz (1987) reported on a strategy first implemented in New York that was aimed at reducing, even eliminating, the incidence of Tay-Sachs. *Chevra Dor Yeshorim* built on the traditions of ultra-Orthodox Judaism, in which potential marriage partners are introduced by a matchmaker. The matchmaker was privy to the results of confidential genetic screening to identify heterozygotes. Boys and girls who were both heterozygous would not be introduced, and all carriers would marry non-carriers. Moreover, since only the matchmaker knew of the genetic status of any marriage candidates, this scheme avoided any stigma of a positive test result[2]. Most importantly, because all babies with Tay-Sachs must have two carrier parents, the frequency of the disease would plummet. In the first four years of operation, in fact, Merz reported no affected births[3]. Defective alleles would seemingly be forever masked in carriers.

The downside of this solution is in the long term, at equilibrium. Campbell (1987) pointed out that, since there are no Tay-Sachs births, there is no selection acting against the recessive allele, which will then increase in frequency because of the continued pressure from de novo mutation. Eventually, an equilibrium will be reached ($q = 0.25$), at which point the masking strategy fails: non-carrier marriage partners will not be able to be found for all carriers. The public-health consequences at this point would be significant, with a far greater incidence of Tay-Sachs than ever before.

But Campbell's doomsday scenario fails to understand the importance of history as approach. Being driven by mutation, the increase in frequency of the defective allele

---

[1] Biology is full of exceptions, of course: individuals with two different disease alleles, who are thus heterozygous, can also have the disease, but these exceptionally rare cases do not concern us here.

[2] The scheme continues today in a slightly different form (see https://doryeshorim.org/faq/) and has been expanded to cover over 50 single-locus recessive disorders. Those tested receive an identification code, but not the information about their test result. Couples contemplating marriage call a 24-hour hotline and submit their codes. Shortly afterwards, their call is returned and both parties are told whether or not they are compatible.

[3] Indeed, the program claims that it "is credited with singlehandedly eradicating Tay Sachs from the Jewish community" (https://doryeshorim.org/our-mission/#history [accessed 6 March 2024]).

is painfully slow. Paul and Spencer (1988) estimated that reaching $q = 0.25$ would take about 1 million years (~40,000 generations at 25 years per generation). By then, of course, human society is likely to be unrecognizable! How we get to equilibrium matters far more than the equilibrium itself.

7.2.3 Mixing Approach and Chance: Mutation-selection Balance in a Finite Population

Of course, more than one flavor of history may be present in our model or study system. Perhaps it is commonplace or even the rule that multiple flavors occur, in fact. Certainly, chance is often found in biological systems and here I want to discuss how it can interact with approach to produce an outcome that would not arise from either flavor acting alone or even with the pair acting additively.

The pioneering population geneticist Sewall Wright first showed how mutation-selection balance and genetic drift interacted, rather than superimposing their separate effects over the top of each other. The latter—additive—outcome would be manifested as drift-induced random fluctuations in the frequency of the mutant allele around the mutation-selection balance equilibrium value. But as Wright (1937) showed, these processes interact to reduce the expected frequency, an effect more pronounced the smaller the population. Consequently, the system does not even approach the deterministic equilibrium of the mutation-selection balance model; rather it wobbles toward a lower value, but seldom reaches it and never stays there.

Fig. 7.1 shows the effect of increasing chance (lower population size) on the mean frequency of a recurring recessive mutation that is lethal when homozygous (i.e., $s = 1$). We can see that for population sizes of close to 1 million, the mean frequency is very close to that predicted by the classical $q = \sqrt{\mu}$ formula. (In general, this formula is fairly accurate for population sizes more than the reciprocal of the mutation rate.) And almost no populations lack the mutant. For population sizes of 1000 or less, however, the majority of populations have no mutants and the mean frequency is less than half $\sqrt{\mu}$. The larger drift-induced fluctuations in mutant frequency occurring in the smaller populations lead to frequent extinction of the mutant, which remains absent for some generations because of the rarity of de novo mutation.

Thus, the deterministic equilibrium value given by the standard formula may never be attained, nor will this value be the long-term average over time. History matters here in two ways: first in that the deterministic equilibrium is approached extremely slowly, and second in that chance in the form of genetic drift interacts with approach so that, on average, the expected frequency is lowered.

This example highlights more generally how important disturbance might be in a system. If a system takes time to return to that equilibrium after a disturbance or is frequently disturbed from its equilibrium, then approach will be an important aspect of its history.

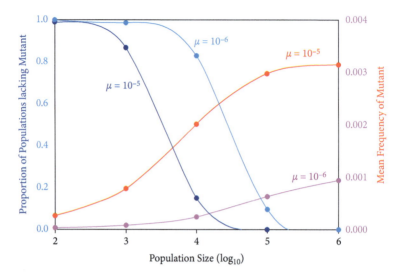

Fig. 7.1 Effect of population size on a recessive lethal for two mutation rates, $\mu = 10^{-5}$ and 10^{-6}. The mean frequency of this mutant is shown by the red and pink lines, plotted against the right-hand axis. In an infinite population, the expected frequency is given by $q = \sqrt{\mu}$, hence ~0.0032 and 0.0010, respectively. The frequency in finite populations will differ among populations: some, particularly small ones, will have no mutants. The expected proportion of populations lacking the mutant is shown by the blue lines, plotted against the left axis.

7.3 Turnover

Perhaps the most fundamental way an equilibrium-centered approach can fail is if there is, in fact, no equilibrium. We have already seen some examples, such as the cycling and chaotic behavior observed in the model of logistic population growth (section 6.3). Although the values around which the population size cycles could be seen as a sort of equilibrium set, it is more difficult to cast chaos in these terms.

The flavor of turnover, however, is slightly different: here the system reaches a steady state, rather than an equilibrium. Superficially, a steady state looks rather like an equilibrium, but there is constant change going on in the system that balances out. Moreover, depending on the exact focus of our interest, different descriptors of the system may be at a steady state or at equilibrium. These alternatives may make it seem as if the difference between equilibrium and steady state is merely semantic; it is not, since the term "steady state" reminds us about the parts of the system that continue to change.

7.3.1 Neutral Theory of Molecular Evolution

The neutral theory of molecular evolution is central to our understanding of population genetics (Kimura, 1983). It holds that most of the genetic variation we see in natural populations—the standing variation—is a balance between the origin of new genetic variants (alleles) that are selectively equivalent to each other and their extinction via genetic drift. Importantly, not all mutations are neutral; most are deleterious and removed through the actions of natural selection. But the variation we commonly observe is continually turning over, different alleles arising, sometimes fixing even, but all eventually drifting to extinction as they are replaced by others. There is some sort of long-term balance in the midst of this transience, however: the long-term average rate at which alleles are fixed (the "substitution rate") is equal to the neutral mutation rate.

Moreover, for a given population size and mutation rate, there is an expected value for heterozygosity (the proportion of heterozygotes in the population):

$$H_E = \frac{4N_e v}{1 + 4N_e v}, \tag{7.1}$$

where N_e is the effective population size and v is the neutral mutation rate. Indeed, this formula is the equilibrium value of the iteration in expected heterozygosity each generation under neutrality, with increases due to mutation and decreases due to drift. (See Box 7.2 for the details of its derivation.)

Interestingly, however, neither the iteration in expected heterozygosity nor its equilibrium value describe at all accurately what goes on in any one population. Fig. 7.2 shows two replicate computer simulations of actual levels of heterozygosity

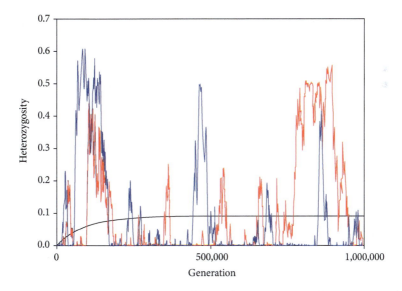

Fig. 7.2 Observed heterozygosity over time in two replicate simulated finite populations (red and blue) under the neutral hypothesis, starting at 0 (monomorphism). Expected heterozygosity (black line) starts at 0.0. In all cases, N_e = 50,000 and $v = 5 \times 10^{-6}$ and hence $H_E \approx 0.0909$. Note that the two observed trajectories are very different; in addition, neither is close to the expected trajectory.

under neutrality in finite populations, compared with the expected value as derived in Box 7.2[4].

> **Box 7.2 Expected Heterozygosity under Neutrality**
>
> Consider a locus subject to neutral mutation at a rate v in a population of effective size N_e. On the one hand, mutation occurring in this population will increase heterozygosity (H) when it occurs in either of the two genes of a homozygote. Thus, on average,
>
> $$H_m = H + 2v(1-H).$$
>
> On the other hand, assuming that the alleles at the locus are selectively neutral, then, on average, the heterozygosity will decrease through drift each generation by a factor of $1/2N_e$. This factor is the probability that the very same gene will be united with a copy of itself in the following generation. Thus, we have an equation for the iteration of expected heterozygosity:
>
> $$H' = H_m \left(1 - \frac{1}{2N_e}\right).$$
>
> Substituting our previous expression for H_m into this latter equation means we can solve $H' = H$ to find the equilibrium value, H_E:
>
> $$H_E = \frac{4N_e v - 2v}{1 + 4N_e v - 2v} \approx \frac{4N_e v}{1 + 4N_e v},$$
>
> the approximation following because $2v$ is so much smaller than the other terms. The eigenvalue of the linearized system evaluated at the equilibrium is $\left(1 - \frac{1}{2N_e}\right)(1 - 2v) \approx 1 - \frac{1}{2N_e} - 2v$, which is very close to one and, hence, the approach to equilibrium is extremely slow.
>
> The *effective number of alleles* is the number of equally frequent alleles required to give the same level of heterozygosity and is given by $A_e = \frac{1}{1-H}$, and so its equilibrium value is $A_E = 1 + 4N_e v$. The above iterations can be cast in terms of A_e or, as in Malécot (1948) and Kimura and Crow (1964), in terms of the homozygosity, $F = 1 - H$.

Both runs of the model exhibit significant fluctuations, with mean values over the 1 million generations ~0.104 and ~0.081, values statistically significantly different from each other and from the expected value of ~0.091. (Of course, such a test violates the most basic of most statistical tests' assumptions, that the data points are independent. There is an element of the same process that leads to capriciousness here.) The expected value started at 0.0 and iterated according to the formula for H' in Box 7.2.

One might think that a model for expected values might mimic fairly closely the actual numbers. In the case of this form of turnover, that is not true, with allele-frequency fluctuations driven largely by genetic drift. Again, history matters,

---

[4] Each generation, the simulated population was subject to mutation, with the number of genes mutating to a novel mutant being a Poisson variable with mean $2N_e v$ (= 0.5). Genetic drift was implemented by taking a random sample with replacement of size $2N_e$ (= 100,000) from the $2N_e$ genes after mutation.

not least because the system is often far away from its long-term average. There is an element of history as chance and history as approach here, too, in addition to history as turnover.

The behavior of heterozygosity under the neutral theory clearly raises the more general question of how well an equilibrium characterizes a system, even in the long term. As in the example of mutation-selection balance (see section 7.2.1), progress of the expected heterozygosity toward the equilibrium is agonizingly slow and hence, once again, the equilibrium may not be a fair reflection of behavior over the long term.

The importance of an historical perspective in understanding the neutral theory was brought home by Nei et al. (1975), who noted that population size often fluctuates. Of particular interest was the case of a population bottleneck, in which the population crashes to a very small size for a few generations before recovering, possibly quite rapidly. Clearly, there ought to be a reduction in heterozygosity, because the smaller population size induces more genetic drift; smaller N_e results in smaller H_E. This scenario has been hypothesized to underlie several theoretical models of speciation (reviewed in Spencer, 1995), as well as the diversity of Hawaiian drosophilids (Carson, 1970).

Nei et al. (1975) asked two important questions: (i) how much heterozygosity is lost in the bottleneck? and (ii) how long does it take for levels of heterozygosity to recover? The short answer is that the details matter: the loss of mean heterozygosity is not too drastic if the population recovers quickly (especially if the minimum population size during the bottleneck is not too small), whereas any reduction that does take place requires millions, possibly hundreds of millions of years to reverse. As these authors pointed out, this recovery period is possibly greater than the evolutionary lifetime of a species, which implies that observed levels of heterozygosity may well be significantly lower than that given by the formula for H_E. In other words, like the predicted equilibrium for a deleterious recessive allele, the equilibrium value for heterozygosity many never be reached.

Nei et al. (1975) also found that the average number of neutral alleles at a locus is drastically reduced by a bottleneck, since many rare alleles drift to extinction. These two different measures of genetic variability—heterozygosity and the number of alleles—give very different pictures of the effects of bottlenecks. Which is more important in evolution was debated in an earlier exchange between Ernst Mayr, who championed the importance of bottlenecks in speciation, and Richard Lewontin, who was skeptical of such models (Lewontin & Mayr, 1965).

The incremental recovery in the value of the expected heterozygosity (H) is, of course, due to the weakness of the process driving it, mutation. Importantly, the effect of genetic drift, which on average reduces H, is likely to be greater in any single population. As the simulations exemplified in Fig. 7.2 reveal, heterozygosity in a single population can depart significantly from the expected heterozygosity. Hence, both the degree of the initial loss in heterozygosity and its post-bottleneck growth seem unlikely to be properly characterized by the deterministic iterations in H. Here history matters, but so too does the temporal scale of investigation.

One might think that the importance of history under neutrality was obvious, especially given Nei et al.'s unequivocal findings and the paper's high number of citations (> 4000). And yet, almost 50 years later, when discussing explanations of the levels of genetic variability observed in natural populations, Charlesworth and Jensen (2022) wrote "[w]e highlight a potentially important role for the less-appreciated contribution of population size change; specifically, the likelihood that many species and populations may be quite far from reaching the relatively high equilibrium diversity values that would be expected given their current census sizes." I couldn't agree more!

7.3.2 Island Biogeography

Robert MacArthur and E. O. Wilson epitomize Kingsland's (1985) ecological theorists, arguing consistently for generalizable models that make robust and measurable ecological predictions, and that suggest possible explanatory factors worthy of further investigation. Nevertheless, the theory of island biogeography they constructed (MacArthur & Wilson, 1963, 1967) is a further example of the role of history as turnover, in spite of their generally anti-historical attitude. The theory was aimed at distilling into a coherent framework the empirical observations that larger islands and those nearer to the mainland have more species of particular groups (e.g., herpetofauna, ants). They hypothesized that the number of species on an island was a balance between the arrival of new colonists and the extinction of existing residents (see Fig. 7.3). Closer islands would have greater immigration rates and larger islands would have lower extinction rates, everything else being equal. MacArthur and Wilson explicitly hoped that their ideas would be soon replaced by stronger and more general theories.

Island biogeography is remarkably similar in structure to Kimura's neutral theory just discussed. The number of species on an island of a certain size is analogous to the effective number of alleles; the arrival of new colonists parallels mutation; and the extinction of residents mimics drift. As the particular alleles constituting the standing variation in the neutral theory are doomed to eventual extinction, even as the effective number fluctuates wildly around an equilibrium, so too are individual species on each island destined for elimination, although in the long term the community remains approximately as diverse. In both cases the composition of the assembly changes—history as turnover—but measures of the variation equilibrate (to some degree).

Nevertheless, there are important differences between island biogeography and the neutral theory that are critical to the history of the model system. In the former, both immigration and extinction rates change as the number of species on the island increases (see Fig. 7.3). Immigration rate declines: with more species on the island, the likelihood that an immigrant is a species not already there decreases. (In the "infinite alleles" model of neutrality, each novel mutation is unique; in island biogeography, the pool of immigrants is the finite set of species found on the mainland.) Extinction, by contrast, becomes commoner: with more species on the island, at any point there are

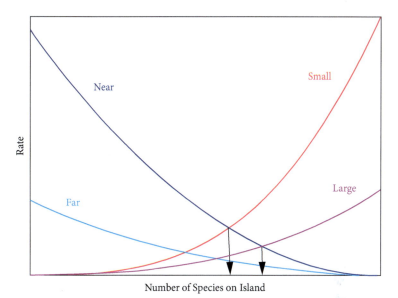

Fig. 7.3 The rates of immigration (blue lines) and extinction (red and pink lines) depend on the number of species on an island. Where these rates are equal determines the number of species present on the island. So, for example, for islands near the mainland (dark blue immigration rate), a small island (red extinction rate) has fewer species than a large island (pink extinction rate).

Modified from MacArthur and Wilson (1963).

more that can die out. Under the neutral theory both mutation and population size were constant over time. (The effect of varying population size in the neutral theory was an important development that came later; Nei et al., 1975.)

Moreover, MacArthur and Wilson (1967) discussed further ways in which the immigration and extinction rates might be modified. For example, if extinction rate were directly proportional to the number of species on an island, the red and pink curves in Fig. 7.3 would be straight lines. They argued, however, that the curves would be concave-up (as shown) because more speciose islands would have lower mean population sizes due to interference and competition among species. Species on such islands would on average, therefore, be more prone to becoming extinct.

Perhaps a more important difference is due to the magnitude of the processes whose balance predicts the number of species on the island. In a field experiment set up to test the theory, both immigration and extinction rates were orders of magnitude greater than typical mutation rates (Simberloff & Wilson, 1969). Consequently, whatever fluctuations occur around the equilibrium number of species are likely to be smaller and shorter-lived than analogous fluctuations in heterozygosity under the neutral theory. Nevertheless, yet again, the variance in the variable of interest, in this case the number of species on an island at time t, was sizeable. The historical perspective that I am advocating makes us aware that the equilibrium (or even the mean) does not capture all that is useful to know.

7.4 Approach and Turnover in Perspective

Both approach and turnover focus on whether we are at an equilibrium or even if one exists. Approach emphasizes the time taken to reach an equilibrium, to the point where the system may only rarely—or even never—be close to equilibrium. In such cases, an equilibrial focus can be positively misleading. Interactions with other flavors of history, such as chance, are likely to be common and can systematically change the long-term behavior of the system. In these cases, the expected value of the variable of interest (e.g., the allele frequency of a deleterious recessive) can be demonstrably different from the predicted equilibrium.

History as turnover reminds us that our systems can be described in different ways: some variables of interest (e.g., heterozygosity) can have an equilibrium expected value, yet parts of the system (e.g., alleles) can be constantly turning over, causing other variables of interest (e.g., allele frequencies) to change. Many systems clearly have an aspect of turnover: individuals in populations change even while population sizes remain constant, and the identity of alleles by descent can change in population-genetic models even while allele frequencies are static. Whether these changes matter or not depends on just what in the system needs to be explained.

Last, I have portrayed turnover above as the arrival and departure (or birth and death) of elements in a system: alleles in the neutral model and species in the theory of island biogeography. But of course, the frequencies of these elements change in between these extremes. There is a degree of turnover even when frequencies of alleles change or when the abundance of different species alters.

References

Campbell R. B. 1987. The effects of genetic screening and assortative mating on lethal recessive-allele frequencies and homozygote incidence. *American Journal of Human Genetics* 41:671–677.
Carson H. L. 1970. Chromosome tracers of the origin of species. *Science* 168:1414–1418.
Charlesworth B., Charlesworth D. 2010. *Elements of Evolutionary Genetics*. Greenwood Village, Col.: Roberts & Co.
Charlesworth B., Jensen J. D. 2022. How can we resolve Lewontin's paradox? *Genome Biology and Evolution* 14:evac096.
Edelstein-Keshet L. 1988. *Mathematical Models in Biology*. Boston, Mass.: McGraw-Hill.
Fisher R. A. 1924. The elimination of mental defect. *Eugenics Review* 16:114–116.
Kimura M. 1983. *The Neutral Theory of Molecular Evolution*. Cambridge, UK: Cambridge University Press.

Kimura M., Crow J. F. 1964. The number of alleles that can be maintained in a finite population. *Genetics* 49:725–738.

Kingsland S. E. 1985. *Modeling Nature: Episodes in the History of Population Ecology*. Chicago: University of Chicago Press.

Lewontin R. C., Mayr E. 1965. Discussion of paper by Dr. Howard. Pages 481–484 in *The Genetics of Colonizing Species*, edited by H. G. Baker and G. L. Stebbins. New York: Academic Press.

MacArthur R. H., Wilson E. O. 1963. An equilibrium theory of insular zoogeography. *Evolution* 17:373–387.

MacArthur R. H., Wilson E. O. 1967. *The Theory of Island Biogeography*. Princeton: Princeton University Press.

Malécot G. 1948. Les Mathématiques de l'Hérédité. Paris: Masson et Cie. Translated into English as *The Mathematics of Heredity* by D. M. Yermanos. 1969. San Francisco: Freeman.

Merz B. 1987. Matchmaking scheme solves Tay-Sachs problem. *Journal of the American Medical Association* 258: 2636–2639.

Nei M., Maruyama T., Chakraborty R. 1975. The bottleneck effect and genetic variability in populations. *Evolution* 29:1–10.

Nielsen R., Slatkin M. 2013. *An Introduction to Population Genetics: Theory and Applications*. Sunderland, Mass.: Sinauer Associates.

Otto S. P., Day T. 2007. *A Biologist's Guide to Mathematical Modeling in Ecology and Evolution*. Princeton, NJ: Princeton University Press.

Paul D. B., Spencer H. G. 1988. Genetic screening and public health. *American Journal of Human Genetics* 43:344–347.

Simberloff D. S., Wilson E. O. 1969. Experimental zoogeography of islands: The colonization of empty islands. *Ecology* 50:278–296.

Spencer H. G. 1995. Models of speciation by founder effect: A review. Pages 141–156 in *Speciation and the Recognition Concept: Theory and Application*, edited by D. M. Lambert and H. G. Spencer. Baltimore: Johns Hopkins University Press.

Wright S. 1937. The distribution of gene frequencies in populations. *Proceedings of the National Academy of Sciences, USA* 23:307–320.

Zuk M., Spencer H. G. 2020. Killing the behavioral zombie: Genes, evolution and why behavior isn't special. *BioScience* 70:515–520.

8

Construction Part 1
Explaining Allelic Diversity

8.1 Construction

Some of the most enduring and fundamental questions in ecology and evolution have to do with variation. For example, Lewontin (1974) argued in his agenda-setting critique, *The Genetic Basis of Evolutionary Change*, that the most fundamental issue in evolutionary genetics was to understand the nature of genetic diversity among organisms, since evolutionary change depends on such variation. This intellectual schema he traced back to his PhD supervisor, Theodosius Dobzhansky (Lewontin et al., 2001). In brief, Lewontin wanted to know how we could explain the levels of genetic variation we observe so commonly in natural populations.

In theoretical ecology, parallel problems about how different species coexist in ecological communities motivated another groundbreaking monograph, *Stability and Complexity in Model Ecosystems* (May, 1973). With species analogous to alleles (or possibly haplotypes), the underlying questions concerned the generation and maintenance of biological diversity that was all but universally observed in natural communities.

In the 1970s both fields were bedeviled with an apparent conundrum. I go into further details below, but in short, analyses of the equilibria of theoretical models seemed to imply that the genetic and ecological diversity found in nature should not exist. This ahistorical approach led writers to conclude that selection was incapable of maintaining more than three or four alleles; highly polymorphic loci revealed by numerous electrophoretic investigations were paradoxical. Similarly, ahistorical equilibrial analyses of models of food-web interactions suggested that we should observe mostly species-poor ecosystems, leaving the rich biodiversity of tropical rain forests and coral reefs unexplained.

The historical approach—history as construction—has, to some degree, resolved this pair of conundrums (which is not to say that all the questions are answered!). Again, I explain in more depth below, but this approach, by adding a temporal dimension to the existing models, allowed diversity to build up over time via, respectively, mutation and species invasion. (I note this parallel between alleles and species also obtains in the comparison of the neutral theory with the theory of island biogeography; see Chapter 7.) So, models of selection, in which novel mutations accumulated over time to produce multi-allelic polymorphisms, revealed a possible explanation of the electrophoretic puzzle, and standard food-web models, in which new species arrived in the system over time, provided a potential answer to the biodiversity question.

I have split my discussion of history as construction into an evolutionary chapter (this one) and an ecological chapter (Chapter 9). My excuse is that in both fields, a significant number of studies have incorporated an historical view and this work has fundamentally changed our understanding of the motivating questions adumbrated above. Perhaps more than for any other flavors, an explicit recognition of history as construction has made a major scientific difference.

This chapter, then, could be entitled, "Homage to Theodosius Dobzhansky, or Why Are There So Many Kinds of Alleles?" The next section takes its title from the central chapter in Lewontin (1974).

8.2 The Paradox of Variation

Population genetics in the 1950s was wracked by arguments about how much genetic variation was present in natural populations. Knowledge of the levels of this standing variation was crucial in understanding evolution, since evolutionary change driven by natural selection relies on genetic differences underpinning the phenotypic differences that are selectively relevant. If standing variation were common, selection could effect change immediately; if not, evolutionary change might have to wait for the relevant mutations to arise. This debate was known as the classical-balance controversy (Beatty, 1987), the position of the so-called classical school being that variation was rare, while the balance school held that variation was ubiquitous.

The advent of electrophoretic data in the 1960s revealed that genetic variation in the form of allozyme polymorphism was abundant in natural populations. And yet, this data somehow failed to resolve matters. Instead, according to Lewontin (1974), the classical and balance schools segued, respectively, into the neutralists and selectionists (see also Hey, 1999), and the debate was transformed to being about the role played by selection in shaping levels and kinds of genetic variation. Instead of the arguments being about whether levels of genetic variation were high (the balance school) or low (the classical school), the dispute became one about whether the levels of variation relevant to selection were high (the selectionists) or low (the neutralists).

The disagreements, then, had the neutralists viewing the standing variation as selectively equivalent, neutral with respect to selection, with the allelic frequencies largely governed by genetic drift, and the selectionists continuing to invoke various forms of balancing selection that actively molded genetic diversity. Lewontin's (1974) survey of population genetics, however, concluded that neither of the explanatory theories, neutralism or selectionism, could adequately explain the electrophoretic data. This mismatch between theory and observation he dubbed the "paradox of variation" (see also Felsenstein, 1975).[1]

8.3 Equilibrium and the Parameter-space Problem

Among the many arguments against the selectionist view was one arising from some of Lewontin's own work, Lewontin et al. (1978). Clearly, this paper postdates his book; the arguments outlined below are additional to those in the original discussion of the paradox. Lewontin et al. (1978) asked a deceptively simple question: What sort of selection would maintain high levels of genetic variability at a single locus? The question came as gel electrophoresis, especially the more sophisticated versions using different conditions (e.g., varying pH or gel materials), was revealing more and more protein polymorphism in a wider and wider range of species from all over the globe. In *Drosophila pseudoobscura*, for instance, Singh et al. (1976) found an astounding minimum number of 37 xanthine dehydrogenase alleles when grinding up just 146 flies.

Lewontin et al. (1978) examined the classical model of constant-viability selection acting on a single locus in an infinite population of diploid organisms with discrete generations. This model, the simplest in a panoply of possible models, had provided the underlying heuristic for the balance school: it had long been known that for two alleles, heterozygote advantage (or heterosis) was both necessary and sufficient to maintain both alleles at a stable equilibrium (see Box 8.1). In Dobzhansky's view, heterosis was the most likely form of balancing selection that would actively maintain variation. Indeed, it remains so today, with an increasing (albeit still small) number of documented cases of heterozygote advantage leading to stable polymorphism (e.g., Gemmell & Slate, 2006; Hedrick, 2012; Hedrick et al., 2014; Strickland et al., 2021; De Pasqual et al., 2022).

---

[1] Confusingly, a number of authors (e.g., Roberts, 2015; Buffalo, 2021; Charlesworth & Jensen, 2022) have more recently used the term "Lewontin's paradox" or even "Lewontin's paradox of variation" to refer to what is effectively just a subset of the original paradox of variation. As part of his overall argument, Lewontin (1974) noted that the neutral theory predicts that heterozygosity (and indeed any measure of genetic diversity) increases with population size: the equilibrium value is predicted to be $H_E = 1 - \frac{1}{4N_e v+1} = \frac{4N_e v}{4N_e v+1}$ (where N_e is the effective population size and v is the neutral mutation rate; see also Chapter 7). Observed heterozygosity values at the time, however, implied that effective population sizes across metazoan species varied by a factor of four or less, a narrow range that seemed patently ridiculous. More recent calculations have slightly different figures, but the conclusion is the same: the neutralist interpretation of genetic diversity predicts too narrow a range of effective population sizes across distantly related organisms. This disagreement was (and is; Kern & Hahn, 2018) seen as strong evidence against neutralist explanations of genetic variation.

Box 8.1 Constant Viability Selection at a Diploid Locus in an Infinite Population with Discrete Generations

This model is a generalization of that outlined in Box 2.2. Suppose we have n alleles, $A_1, A_2, \ldots, A_n$ at respective frequencies $p_1, p_2, \ldots, p_n > 0$ (and so $\sum_{i=1}^{n} p_i = 1$). Let w_{ij} (= w_{ji}) be the (constant) viability of genotype $A_i A_j$ and then $\bar{w} = \sum_{i=1}^{n} \sum_{j=1}^{n} w_{ij} p_i p_j$ is the mean fitness (viability) of the population. Allele frequencies iterate each generation according to
$$p_i' = p_i \sum_{j=1}^{n} w_{ij} p_j / \bar{w}.$$

It can be shown (Kingman, 1961) that $\bar{w}$ is a non-decreasing function over time: the mean fitness increases until the system reaches an equilibrium (at which $p_i' = p_i$ for all i). This finding matches our intuitive understanding of how selection leads to fitness-maximizing adaptation. A fully polymorphic equilibrium exists when the equilibrium allele frequencies, $\hat{p}_i$, are all non-zero. At most one such equilibrium can exist for a given set of viabilities and, if it is stable, it is globally stable, reached by all initial sets of non-zero allele frequencies (Kingman, 1961).

For $n = 2$, the polymorphic equilibrium is (globally) stable if and only if the heterozygote, $A_1 A_2$, is fitter than both homozygotes, $A_1 A_1$ and $A_2 A_2$. This condition, $w_{12} > w_{11}, w_{22}$, is known as heterozygote advantage or heterosis.

For larger values of n, however, the situation is not at all clear. Although there is a precise analytical condition for the selective maintenance of three alleles, it does not correspond to an intuitive extension of the two-allele condition of heterozygote advantage. Lewontin et al. (1978) provide counterexamples to both the necessary ("only if") and sufficient ("if") aspects. These authors exhibit, respectively, a set of viabilities for three alleles without heterozygote advantage that nevertheless leads to a stable three-allele equilibrium (so heterozygote advantage is not necessary), as well as a case in which all three possible heterozygote viabilities are greater than the viabilities of all three homozygotes and yet there is no three-allele polymorphism (and so heterozygote advantage is not sufficient).

In part, this complication arises because the number of genotypes (and thus viabilities) increases rapidly, and there are too many values to combine into a simple heuristic explanation. In general, there are n homozygotes and $\binom{n}{2}$ heterozygotes, a total of $n + \binom{n}{2} = \frac{2n}{2} + \frac{n(n-1)}{2} = \frac{n(n+1)}{2}$ genotypes and hence viability parameters.

For more than two alleles, however, there is no simple intuitive condition for polymorphism (see Box 8.1). Thus, even for the simplest model of selection, it was not obvious what was required to maintain higher levels of variation, let alone whether selection could, more generally, do so.

Lewontin et al. (1978) approached this problem by asking, instead, what proportion of parameter space ("fitness space") allowed for a stable full polymorphism. In other words, given random viabilities for an n-allele model, what was the probability the model would exhibit a stable equilibrium with all n alleles present? Everything hinges on the word "random." They sampled their viability values independently from a uniform distribution between zero and one. Both independence and uniformity are critical. Independence means that the fitness of A_2A_3 is independent of that of, for example, A_2A_4; there is no effect of A_2 alone. Uniformity means that all values are equally likely. I return to these assumptions and what happens when they are relaxed below.

For $n = 2$, the answer to Lewontin et al.'s question is straightforward. There are three possible genotypes and hence three viabilities; the condition for a stable two-allele polymorphism is that the heterozygote fitness is the largest of these three values, which happens a third of the time when these values are randomly and independently chosen. But, for the same reasons as we saw above, for $n > 2$, there is no simple answer.

Lewontin et al. resorted to computer simulation, sampling 10^5 random sets of the required viabilities for each value of n. Each set of viabilities was analyzed to determine: (i) whether it afforded an equilibrium with all n alleles present (i.e., there was a "feasible" equilibrium) and (ii) whether that equilibrium was stable. Positive answers to both (i) and (ii) meant that that set of fitnesses led to a stable n-allele polymorphism; the answer to their question was simply the number of these fitness sets divided by 10^5. Their results are summarized in Fig. 8.1.

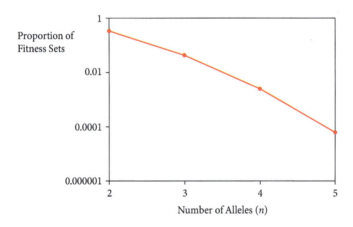

Fig. 8.1 The proportion of random n-allele fitness sets leading to a stable, full polymorphism (i.e., stable equilibria with n alleles). Note the log scale of the y-axis.

Data from Lewontin et al. (1978).

As n increased, the proportion of their fitness sets maintaining all the variation dropped precipitously. Indeed, for $n = 6$, they could not find a single set of random fitnesses among the 10,000 they generated that maintained all six alleles. Clearly, if constant viability selection were maintaining the levels of variation found in Singh et al. (1976), the selection parameters must be exactly special, coming from an incredibly small part of parameter space.

It is worth pointing out that a dramatic decline in the fraction of fitness space affording a full polymorphism appears to be fairly robust to changes in the underlying population-genetic model. Lewontin et al. (1978) performed further simulations with some restrictions on the fitnesses (e.g., requiring "pairwise heterosis;" see Box 8.2) but found the same qualitative outcome. The same result occurred in models with fertility selection (Clark & Feldman, 1986), sex-specific selection (Marks & Ptak, 2000), maternal selection (Spencer & Chiew, 2015) and frequency-dependent selection (Trotter & Spencer, 2007), as well as models with different mutational fitness distributions (Trotter & Spencer, 2013; Spencer & Mitchell, 2016) and spatial structure (Star et al., 2007a).

Lewontin et al. (1978) were careful about the interpretation of these results, writing, "we are *not* trying to make statements about the 'probability' that stable equilibria will occur" (page 152). Nobody thinks, of course, that real viabilities are drawn independently from a uniform distribution. A statement about probability would require knowledge (or assumptions) about the distribution of fitnesses of newly arising mutations in the real world (yet another vexed question in evolutionary genetics). Nevertheless, their method does characterize viability-parameter space in some coherent manner.

> **Box 8.2 Heterozygote Advantage with More than Two Alleles**
>
> With two alleles, it is quite clear what we mean by heterosis (or heterozygote advantage). Continuing with our terminology from Box 8.1, we require $w_{12} > w_{11}, w_{22}$. But with three alleles, there are at least two potential definitions. Using Lewontin et al.'s terminology, "pairwise heterosis" entails that $w_{ij} > w_{ii}, w_{jj}$ for all $i \neq j$: each heterozygote is fitter than the two corresponding homozygotes. "Total heterosis," by contrast, is a stronger condition, requiring that all heterozygotes have larger viabilities than all homozygotes: $\min_{i \neq j}(w_{ij}) > \max_{k}(w_{kk})$. A further possibility is what one might call "average heterosis:" $w_{ij} > \frac{1}{2}(w_{ii} + w_{jj})$ for all $i \neq j$. In fact, Lewontin et al. (1978) showed that this last condition is necessary for a stable equilibrium in their model, and hence so is $\bar{w}_{ij} > \bar{w}_{kk}$, where the bar indicates the unweighted arithmetic mean.
>
> In Lewontin et al.'s (1978) original set of simulations, the results of which are shown in Fig. 8.1, there was no guarantee that a set of viabilities affording a stable full polymorphism satisfied the requirements for either pairwise or total heterosis. Indeed, given that neither property is a requirement for a stable full polymorphism, it is highly likely that some, possibly even most, of their variation-maintaining fitness sets did not. They did carry out further simulations restricting the viabilities to being pairwise or totally heterotic. Although these fitness sets were better at maintaining variation, there was the same trend of a drastically declining ability as n increased.

Subsequent interpretations of these results, however, have not been so careful. Even in their own paper, Lewontin et al. (1978) forgot their own admonition, concluding, "heterosis alone is not a mechanism for maintaining many alleles" (p. 149). Kimura (1983: 282) cited the paper as part of his argument against the selectionist school as showing "that heterozygote superiority in fitness alone can not maintain many alleles." Many others, even quite recently, have made the same interpretation (e.g., Charlesworth, 2006; Gloria-Soria et al., 2012; Ejsmond et al., 2014; Radwan et al., 2020).

Such inferences are, of course, a logical fallacy. Quite generally, the magnitude of a particular fraction of parameter space tells us almost nothing about the likelihood that real parameters actually occur there. Incorporating some history, as we will see below, gives us a very different view although, ironically, it downplays heterosis per se.

8.4 Constructing Polymorphisms

The idea that genetic polymorphisms maintained by selection may be constructed over time, via a natural process that included mutation, was inspired by the analogous problem in theoretical ecology (see Chapter 9; Taylor, 1988). Again, alleles correspond to species and mutation to immigration.

In their incorporation of an historical perspective, Spencer and Marks (1988; Marks & Spencer, 1991) took the same n-allele viability selection scheme as Lewontin et al. (1978), but added a temporal dimension that allowed novel mutations to bombard the existing population each generation. Starting their simulations with just one allele (usually with homozygous fitness 0.5), they tracked the number of alleles in the population (n) over many generations.

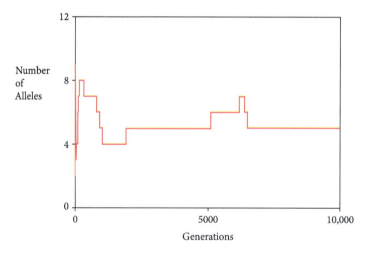

Fig. 8.2 The number of alleles (n) over time, starting with $n = 1$ in a typical simulation run of Spencer and Marks (1988). The number of alleles very rapidly builds up to $n \geq 4$, values that would be exceedingly unlikely to maintain polymorphism if fitnesses had been chosen randomly.

Each generation, a novel mutation arose, adding to the n alleles already present. In keeping with Lewontin et al.'s (1978) approach, their simulation study sampled the viabilities of the $n + 1$ newly possible genotypes (i.e., those involving the new mutant) independently from the uniform distribution from zero to one. As a result of selection, most mutations immediately declined in frequency and became extinct, failing to join the existing set of alleles. Occasionally, however, an allele would succeed and the level of diversity would increase. Sometimes, too, an existing allele would be driven to extinction as others increased in frequency. A typical run of their model can be seen in Fig. 8.2.

Fig. 8.2 shows that there was an initial rapid increase in the number of alleles, with subsequent extinctions, but change decelerated and after ~6500 generations (until 10,000 generations when the simulation was terminated) there were just five alleles present. Fig. 8.3 summarizes the end results of a thousand of such simulations, recording the number of alleles after 10^4 generations. Not dissimilar results have been obtained in long-term experimental evolution studies (e.g., Tenaillon et al., 2016).

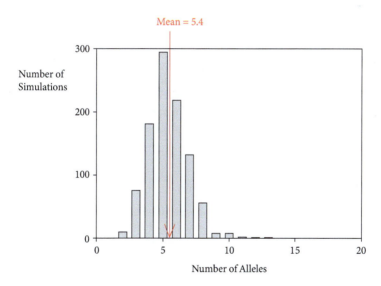

Fig. 8.3 The number of constructionist simulations (out of 1000) with n alleles at Generation 10,000.

After Marks and Spencer (1991).

What is immediately obvious is that the number of alleles present after 10,000 generations is significantly greater than might be expected from a naive interpretation of Lewontin et al. (1978). The latter found just six out of 10^5 simulations were able to maintain all five alleles (and none maintained six), whereas > 43% of the current simulations finished with six or more alleles. Clearly, the process in the Spencer and

Marks (1988) simulations is adept at finding those minute parts of parameter space that maintain larger numbers of alleles.

There are several caveats that apply here, however. First, there is no assurance that heterosis is maintaining the variation; all that we know is that polymorphism is present. This caveat also applies to Lewontin et al.'s simulations, of course (see Box 8.2 for more detail on this matter). Second, we do not know for certain that the systems at Generation 10,000 are at equilibrium. (See Box 8.3 for further details here.) Nevertheless, the behavior seen in Fig. 8.2, where the number of alleles does not change for thousands of generations, suggests that inferring equilibrium in the vast majority of cases isn't unreasonable. Consequently, it seems a safe conclusion that simple viability selection, if not heterosis, could, in principle, actively maintain many alleles at equilibrium.

One further advantage of the historical approach is that it emphasizes how the set of parameter values in the system (i.e., fitnesses in the population) are very different from the universe from which they were chosen. The alleles that succeed in establishing themselves in the population are a distinctly non-random subset of the overall set of mutants; the viabilities of the genotypes in the population are by no means uniform between zero and one. It could be said that the fitnesses themselves—not just the alleles—have evolved (Marks & Spencer, 1992). They have "emergent properties" undetermined at the start of the simulations.

Perhaps most importantly, however, we need to realize that, just as the Lewontin et al. (1978) study does not logically rule out the selective maintenance of allelic variation, the Spencer and Marks (1988) "constructionist" approach does not show that it actually occurs. All we have essentially is an evolution-like process that can generate such an outcome; it is plausible that selection might maintain variation.

> **Box 8.3 How Important is the Equilibrium?**
>
> A natural question to ask is if the simulated populations constructed by Spencer and Marks (1988) (and the various others who have worked on these models) had reached an equilibrium by the time they were terminated. The short answer is that we do not know for sure, although the long period without a change in allele number, as shown in Fig. 8.2, suggests that that particular simulation had done so. Such a pattern was common in their simulations, implying that many if not most populations were, indeed, at an equilibrium, at least with respect to numbers of alleles.
>
> Moreover, a second set of simulations reported in the same paper used a slightly different temporal scale. After each successful invasion by a novel mutant (i.e., when it increased in frequency), no more mutations occurred and the system iterated to equilibrium before mutation was restarted. In this set of simulations, therefore, equilibrium was certainly attained after each invasion. Importantly, the number of alleles at the end of these runs was not noticeably different.
>
> Perhaps hidden in this question, though, is a view that being at equilibrium is what we are interested in. Is that truly the case? We have no guarantee that the levels of variation seen

> **Box 8.3** *Continued*
>
> in natural populations are at equilibrium. Indeed, it seems to me that whether or not a particular population has reached an equilibrium level of diversity is an interesting empirical question.
>
> Nevertheless, it must be acknowledged that the question has changed somewhat. Instead of being about whether or not selection *maintains* genetic variation, it is now about whether selective differences *allow* genetic variation. If the genetic makeup of the population is in flux (over long periods, possibly), selection is merely preserving the level of variation, not the variation itself. History as construction has somehow become history as turnover; equilibrium has become steady state.

Nevertheless, this degree of plausibility has been enhanced somewhat by subsequent studies that relax or change many of the assumptions in the original Spencer and Marks (1988) models. Different forms of selection such as frequency-dependent selection (Trotter & Spencer, 2008, 2013), sex-specific selection (Marks & Ptak, 2000) and maternal selection (Spencer & Chiew, 2015), as well as changing the distribution of mutational fitnesses (Spencer & Marks, 1992; Trotter & Spencer, 2013; Spencer & Mitchell, 2016), introducing population structure (Star et al., 2007b, 2008) and incorporating genetic drift (Star & Spencer, 2013; Spencer & Mitchell, 2016) all give qualitatively similar results: polymorphism builds up easily over time, as selection sifts through the newly arising mutants to preserve those with extremely particular, polymorphism-maintaining fitnesses.

8.5 The Paradox Remains

In spite of these broad findings, however, the constructionist simulations do not completely solve the paradox of variation. For a start, the simpler models (especially the original constant-viability models of Spencer and Marks, 1988 and Marks and Spencer, 1991) generated lots of heterozygote advantage. Perhaps unsurprisingly, if new mutants arose that, in conjunction with alleles already present in the population, exhibited higher fitness as a heterozygote, they were likely to be preserved and added to the existing standing variation. Moreover, the fitness of a mutant homozygote is almost irrelevant to its probability of successful invasion (since the homozygote will initially be rare). Thus, the simulations were always likely to generate significant levels of heterozygote advantage. But heterozygote advantage is, as has already been mentioned, rare in natural populations. Hence, these models have yet to properly capture what is going on in nature.

A related problem, first noted by Spencer and Marks (1993), is that nearly all the simulations predicted too much polymorphism, in contrast to what might have been expected from the Lewontin et al. (1978) study. Rarely, if ever, did the constructionist approach predict a locus without variation, and yet, monomorphic loci are not uncommon in natural populations. At the same time, the simulations largely failed to predict larger numbers of alleles. Yet again, the constructionist models do not adequately match the real world.

> **Box 8.4 Relaxing the Independence Assumption: Models of Generalized Dominance**
>
> Sampling the required viabilities independently from the uniform (or other) distribution means that an allele with a low fitness when homozygous is just as likely to have a high heterozygote fitness when in combination with another allele as an allele that has high homozygous fitness. There is no overall allelic effect. Such an assumption seems implausible given that gene products often derive from the genetic coding on a single chromosome. The biology thus suggests that the fitness of A_iA_j types in the population may be the average of the fitness contribution of each allele (or some value close to this).
>
> To avoid the independence assumption of the previous constructionist models, Spencer and Mitchell (2016) introduced the concept of generalized dominance, in which fitnesses were functions of the individual alleles comprising the genotype plus some, usually lesser weighted, interaction term.
>
> Each allele, A_i, was characterized by a primary effect, X_i, each value independently drawn from the uniform distribution on [0,1]. Each pair of alleles, A_i and A_j, was also given an interaction effect, Y_{ij}, independently sampled from the same distribution. The fitness of genotype A_iA_j was then specified by the weighted sum
>
> $$w_{ij} = \alpha\left(X_i + X_j\right) + (1 - 2\alpha)\, Y_{ij}$$
>
> in which the weight, α, was a fixed constant between 0 and ½. When $\alpha = 0$, all w_{ij} are independent; when $\alpha = $ ½, the fitness of a genotype is simply the average of its component primary effects and heterozygote advantage is impossible. The correlation between the fitnesses of types sharing one allele is
>
> $$\mathrm{Cor}\left(w_{ij}, w_{ik}\right) = \frac{\alpha^2}{1 - 4\alpha + 6\alpha^2},$$
>
> which is a strictly increasing function for $0 \leq \alpha \leq $ ½ (Spencer & Mitchell, 2016).

Perhaps the "generalized dominance" models of Spencer and Mitchell (2016) do the best job of matching real data. Their models dispensed with the idea that fitnesses were independent by assuming, instead, that each allele had some primary fitness effect affecting the viability of all genotypes in which it was present (see Box 8.4). Greater weighting of these primary effects led to a reduction in the numbers of alleles compared with previous studies (Fig. 8.4, black bars). The addition of genetic drift led to some simulations manifesting monomorphism when terminated and others showing five or more alleles, even in finite populations (see Fig. 8.4, blue and gray bars). Still, we're not there yet: what we need is to explain monomorphism and, simultaneously, the 37 alleles of Singh et al. (1976) and even the increasingly vast numbers at MHC loci (Stefan et al., 2019).

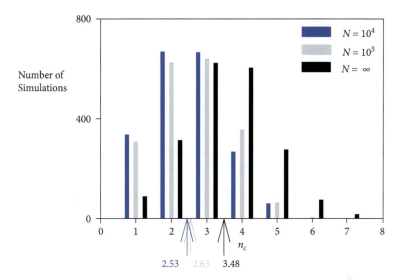

Fig. 8.4 The number of generalized-dominance constructionist simulations (out of 2000) with n_c common alleles (i.e., those at frequencies of 0.01 or greater) at Generation 10,000.

After Spencer and Mitchell (2016).

8.6 Evolutionary Construction in Perspective

Taking an historical approach in the investigation of the ability of selection to maintain allele variation has led to a far more nuanced understanding of the role of selection in molding genetic diversity on natural populations. The ahistorical studies of the properties of selection-parameter space revealed that the proportion of that space affording polymorphisms is minute for more than four or five alleles; the historical, constructionist investigations showed that these very special parts of parameter space could be easily reached by a process of mutation and selection over time. The former finding, while important and interesting (especially in light of the latter result), does not mean that selection cannot maintain allelic variation. The latter finding, too, has its limitations: it does not show that selection does maintain such variation, only that it may plausibly do so.

References

Beatty J. 1987. Weighing the risks: Stalemate in the Classical/Balance controversy. *Journal of the History of Biology* 20:289–319.

Buffalo V. 2021. Quantifying the relationship between genetic diversity and population size suggests natural selection cannot explain Lewontin's paradox. *eLife* 10:e67509.

Charlesworth B., Jensen J. D. 2022. How can we resolve Lewontin's paradox? *Genome Biology and Evolution* 14:evac096.

Charlesworth D. 2006. Balancing selection and its effects on sequences in nearby genome regions. *PLoS Genetics* 2:e64.

Clark A. G., Feldman M. W. 1986. A numerical simulation of the one-locus, multiple-allele fertility model. *Genetics* 113:161–176.

De Pasqual C., Suisto K., Kirvesoja J., Gordon S., Ketola T., Mappes J. 2022. Heterozygote advantage and pleiotropy contribute to intraspecific color trait variability. *Evolution* 76:2389–2403.

Ejsmond M. J., Radwan J., Wilson A. B. 2014. Sexual selection and the evolutionary dynamics of the major histocompatibility complex. *Proceedings of the Royal Society of London B* 281:20141662.

Felsenstein J. 1975. Review of *The Genetic Basis of Evolutionary Change*. *Evolution* 29:587–590.

Gemmell N. J., Slate J. 2006. Heterozygote advantage for fecundity. *PLoS ONE* 1:e125.

Gloria-Soria A., Moreno M. A., Yund P. O., Lakkis F. G., Dellaporta S. L., Buss L. W. 2012. Evolutionary genetics of the hydroid allodeterminant *alr2*. *Molecular Biology and Evolution* 29:3921–3932.

Hedrick P. W. 2012. What is the evidence for heterozygote advantage selection? *Trends in Ecology & Evolution* 27:698–704.

Hedrick P. W., Stahler D. R., Dekker D. 2014. Heterozygote advantage in a finite population: Black color in wolves. *Journal of Heredity* 105:457–465.

Hey J. 1999. The neutralist, the fly and the selectionist. *Trends in Ecology & Evolution* 14:35–38.

Kern A. D., Hahn M. W. 2018. The neutral theory in light of natural selection. *Molecular Biology and Evolution* 35:1366–1371.

Kimura M. 1983. *The Neutral Theory of Molecular Evolution*. Cambridge, UK: Cambridge University Press.

Kingman J. F. C. 1961. A mathematical problem in population genetics. *Mathematical Proceedings of the Cambridge Philosophical Society* 57:574–582.

Lewontin R. C. 1974. *The Genetic Basis of Evolutionary Change*. New York: Columbia University Press.

Lewontin R. C., Ginzburg L. R., Tuljapurkur S. D. 1978. Heterosis as an explanation for large amounts of polymorphism. *Genetics* 88:149–170.

Lewontin R. C., Paul D., Beatty J., Krimbas C. B. 2001. Interview of R. C. Lewontin. Pages 22–61 in *Thinking about Evolution: Historical, Philosophical, and Political Perspectives*, edited by R. S. Singh, C. B. Krimbas, D. B. Paul and J. Beatty. Cambridge, UK: Cambridge University Press.

Malécot G. 1948. Les Mathématiques de l'Hérédité. Paris: Masson et Cie. Translated into English as *The Mathematics of Heredity* by D. M. Yermanos. 1969. San Francisco: Freeman.

Marks R. W., Ptak E. E. 2000. The maintenance of single-locus polymorphism. V. Sex-dependent viabilities. *Selection* 1:217–228.

Marks R. W., Spencer H. G. 1991. The maintenance of single-locus polymorphism. II. The evolution of fitnesses and allele frequencies. *American Naturalist* 138:1354–1371.

May R. M. 1973. *Complexity and Stability in Model Ecosystems*. Princeton: Princeton University Press.

Radwan J., Babik W., Kaufman J., Lenz T. L., Winternitz J. 2020. Advances in the evolutionary understanding of MHC polymorphism. *Trends in Genetics* 36:298–311.

Roberts R. G. 2015. Lewontin's paradox resolved? In larger populations, stronger selection erases more diversity. *PLoS Biology* 13:e1002113.

Singh R., Lewontin R. C., Felton A. A. 1976. Genetic heterogeneity within electrophoretic "alleles" of xanthine dehydrogenase in *Drosophila pseudoobscura*. *Genetics* 84:609–629.

Spencer H. G., Chiew K. X. 2015. The maintenance of single-locus polymorphism by maternal selection. *G3: Genes, Genomes, Genetics* 5:963–969.

Spencer H. G., Marks R. W. 1988. The maintenance of single-locus polymorphism. I. Numerical studies of a viability selection model. *Genetics* 120:605–613.

Spencer H. G., Marks R. W. 1992. The maintenance of single-locus polymorphism. IV. Models with mutation from existing alleles. *Genetics* 130:211–221.

Spencer H. G., Marks R. W. 1993. The evolutionary construction of molecular polymorphisms. *New Zealand Journal of Botany* 31:249–256.

Spencer H. G., Mitchell C. 2016. The selective maintenance of allelic variation under generalized dominance. G3: Genes, Genomes, *Genetics* 6:3725–3732.

Star B., Spencer H. G. 2013. Effects of genetic drift and gene flow on the selective maintenance of genetic variation. *Genetics* 194:235–244.

Star B., Stoffels R. J., Spencer H. G. 2007a. Single-locus polymorphism in a heterogeneous two-deme model. *Genetics* 176:1625–1633.

Star B., Stoffels R. J., Spencer H. G. 2007b. Evolution of fitnesses and allele frequencies in a population with spatially heterogeneous selection pressures. *Genetics* 177:1743–1751.

Star B., Trotter M. V., Spencer H. G. 2008. Evolution of fitnesses in structured populations with correlated environments. *Genetics* 179:1469–1478.

Stefan T., Matthews L., Prada J. M., Mair C., Reeve R., Stear M. J. 2019. Divergent allele advantage provides a quantitative model for maintaining alleles with a wide range of intrinsic merits. *Genetics* 212:553–564.

Strickland L. R., Fuller R. C., Windsor D., Cáceres C. E. 2021. A potential role for overdominance in the maintenance of colour variation in the Neotropical tortoise beetle, *Chelymorpha alternans*. *Journal of Evolutionary Biology* 34:779–791.

Taylor P. J. 1988. The construction and turnover of complex community models having generalized Lotka-Volterra dynamics. *Journal of Theoretical Biology* 135:569–588.

Tenaillon O., Barrick J. E., Ribeck N., Deatherage D. E., Blanchard J. L. et al. 2016. Tempo and mode of genome evolution in a 50,000-generation experiment. *Nature* 536:165–170.

Trotter M. V., Spencer H. G. 2007. Frequency-dependent selection and the maintenance of genetic variation: Exploring the parameter space of the multiallelic pairwise interaction model. *Genetics* 176:1729–1740.

Trotter M. V., Spencer H. G. 2008. The generation and maintenance of genetic variation by frequency-dependent selection: Constructing polymorphisms under the pairwise interaction model. *Genetics* 180:1547–1557.

Trotter M. V., Spencer H. G. 2013. Models of frequency-dependent selection with mutation from parental alleles. *Genetics* 195:231–242.

9

Construction Part 2
Model Ecosystems and Theoretical Ecology

9.1 The Ecological Paradox

We saw in the previous chapter how the introduction of an historical dimension partially solved a long-standing mismatch between observation and theory in population genetics. This matter is paralleled quite remarkably in theoretical ecology, although the terminology was different. Taylor (1989), for example, originally described the ahistorical approach as "morphological" and the historical approach as "developmental;" he subsequently used the word "constructionist" for the latter (Taylor, 2005: 10; Taylor, 2018). Moreover, the realization that complexity arises because of a selective process over time was present in the ecological debate right at the start, although the implications of this recognition were underappreciated.

The ecological paradox arose when theoretical modeling contradicted population ecologists' empirical data about the relationship between trophic complexity and stability of ecosystems. Based on the apparent resilience of highly speciose ecosystems like tropical rain forests and coral reefs to various sorts of perturbations wrought, for example, by storm events, ecologists had surmised that stability was positively correlated with complexity (Elton, 1958). Similarly, simpler ecological communities, especially those on islands or artificial agricultural systems, seemed vulnerable to disturbances, with frequent species extinction and even collapse of the whole community being possible (Elton, 1958). The conclusion that complexity should lead to stability was even intuitively reasonable. And there appeared even to be support from food-web modeling (MacArthur, 1955). There were some doubters and possible counterexamples, but these were few and viewed as exceptional.

Further theoretical modeling, however, spectacularly failed to support this idea. Indeed, as Robert May (1973) discussed in detail, model ecosystems with more tropic links were *less* stable than simple ones. May deliberately examined the simplest of mathematical models of predator-prey systems (Lotka-Volterra equations; see Box 9.1 and Fig. 9.1), since ecologists had previously invoked the lack of a stable equilibrium (and, instead, oscillations in population size) in the single-predator, single-prey version as evidence in line with their prediction (i.e., that complexity implies stability).

If such a simple two-taxa model did not lead to a stable equilibrium, the argument ran, then stability must require more species, more complexity.

But, as May (1973) pointed out, the fluctuating population sizes of the single predator and prey models are only evidence in favor of complexity leading to stability if more speciose models (with more trophic links) displayed less of this non-equilibrial behavior. But they do not. In fact, for $n > 1$, an n-predator, n-prey system is at best only as stable as one with $n = 1$; in general, it is unstable to the point where one or more species becomes extinct. Thus, the paradox: empirical observations suggested more complex ecosystems were more stable, but more complex theoretical models of biological communities were, apparently, less stable.

Box 9.1 Lotka-Volterra Equations for One Predator and One Prey

This model, one of the oldest in population biology, concerns the population sizes of two species (the predator and the prey) in continuous time. It assumes that, in the absence of a predator, prey population size will grow exponentially, but that predator numbers counteract this increase. Conversely, in the absence of prey, predator population size declines exponentially, but this decline is opposed by greater numbers of prey. Following May (1973), let $H(t)$ and $P(t)$ be the population sizes of prey and predator, respectively, at time t. Then, if a, b, α and β are all positive, parametrizing, respectively, prey birth rate, predator death rate, the effect of predators on prey and vice versa, we have

$$\frac{dH(t)}{dt} = H(t)\left[a - \alpha P(t)\right]$$

$$\frac{dP(t)}{dt} = P(t)\left[-b + \beta H(t)\right].$$

This system is at an equilibrium, $(\hat{H}, \hat{P})$, when both these equations equal zero, which immediately gives $\hat{H} = b/\beta$ and $\hat{P} = a/\alpha$. We would like to know the conditions for stability, which means determining when $H(t) \to \hat{H}$ and $P(t) \to \hat{P}$.

Doing so requires some simple calculus and linear algebra (see Roughgarden, 1979: 560; Otto & Day, 2007: 254), to find the linearization around $(\hat{H}, \hat{P})$. By letting $h(t) = H(t) - \hat{H}$ and $p(t) = P(t) - \hat{P}$, we can derive (see May, 1973: 188) this approximation (writing it in matrix form, with the prime indication differentiation with respect to t) to be

$$\begin{pmatrix} h'(t) \\ p'(t) \end{pmatrix} \approx \begin{pmatrix} 0 & -\alpha b/\beta \\ \beta a/\alpha & 0 \end{pmatrix} \begin{pmatrix} h(t) \\ p(t) \end{pmatrix}.$$

The eigenvalues of this matrix are $\pm i\sqrt{ab}$; the non-zero imaginary part means that the values of $H(t)$ and $P(t)$ plotted on a graph with axes for H and P rotate around the equilibrium and the zero real part means that this rotation continues ad infinitum, as cycling, retracing the same values forever. In other words, the equilibrium is neutrally stable. In terms of predator and prey numbers this behavior translates into sustained oscillations that neither dampen nor grow. (See Fig. 9.1 for an illustration.)

110 • Beyond Equilibria

Box 9.1 *Continued*

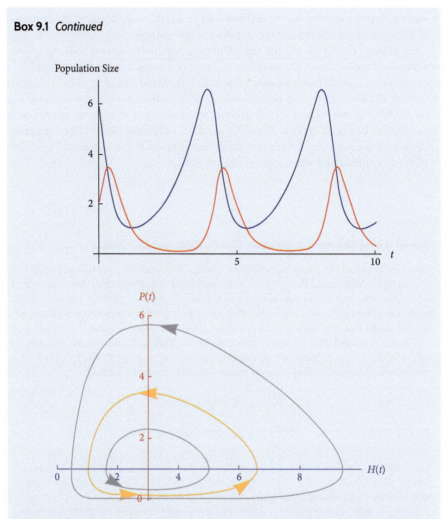

Fig. 9.1 A numerical example of population sizes governed by the Lotka-Volterra equations above (with $a = \alpha = \beta = 1$ and $b = 3$). The unique equilibrium is given by $\hat{H} = 3$, $\hat{P} = 1$. The upper graph shows the population sizes over time of the prey (blue; $H(t)$) and predator (red; $P(t)$), given initial conditions $H(0) = 6$ and $P(0) = 2$. Note that the system oscillates without approaching (or moving away from) the equilibrium. The lower phase-plot graph shows how $H(t)$) and $P(t)$ co-vary over time, depending on the initial conditions (history as contingency). The orange curve corresponds to the upper graph.

These questions, it will be noticed, are all concerned with equilibria and the number of species present therein. There is no historical dimension. This chapter sets out to show how adding the historical flavor of construction to ecological modeling has provided a productive way forward in our understanding of complexity in model ecosystems. Such ideas have also influenced empirical ecology, inspiring studies like that by Stachowicz et al. (1999) on the invasibility of experimental marine communities with different numbers of resident species.

9.2 Recognizing the Ecological Paradox

Ecologists' interest in the stability of biological communities stems from a long-standing desire to explain the persistence of such systems over considerable periods of time. More fundamental, however, was the goal of understanding why these communities are so diverse (and occasionally, why they are not more so). As Yale ecologist George Evelyn Hutchinson (1959) asked in his highly influential presidential address[1] to the American Society of Naturalists, "Why are there so many kinds of animals?"

The first theoretical work in this field was carried out in a very general context, concerned with the issue of the stability of large systems, not necessarily biological, let alone ecological. Gardner and Ashby (1970), for instance, examined the linear system given by the matrix equation $x' = Ax$, where x was the vector of state variables $(x_1, x_2, \ldots, x_n)^T$. The elements of the $n \times n$ matrix, A, were chosen at random given several constraints, most notably restricting just a small proportion, C, known as the "connectance," to be non-zero[2]. The prime indicates the iterated values of x. Gardner and Ashby (1970) generated large numbers of random A matrices for different values of n and C, testing the equilibrium of each corresponding equation for stability. In brief, they found that larger values of n and C both lead to a lower fraction of matrices affording stable equilibria.

May (1972) extended and confirmed these conclusions, and applied the ideas to theoretical ecology. The state variables (the x_i values) were the population sizes of the different species near an equilibrium and the elements of matrix A summarized the intra- and inter-species interactions of the ecological community. Connectance was a measure of the proportion of species that interacted with each other. On the basis of these results, May (1972) argued that stable multi-species communities should consist of species with few, mostly weak, interactions. Such a prediction appeared to fly in the face of Hutchinson's (1959: 155) claim "that the reason why there are so many species of animals is at least partly because a complex trophic organization of a community is more stable than a simple one."

---

[1] With over 5000 citations, dozens of scientific articles whose own titles derive from Hutchinson's original and at least one monograph that also does so, this paper, titled "Homage to Santa Rosalia," has made an amazing impact. Beyond science, even, it is plausible that it has left a footprint in Kurt Vonnegut's science-fiction novel, *Galapagos* (Marshall, 2017). Hutchinson's broad influence over much of ecological research has led to him being called the "Father of Modern Ecology."

[2] The diagonal elements were sampled uniformly from [−1.0, −0.1]; off-diagonal elements were zero with probability 1 − C, and the non-zero values were drawn from the uniform distribution on [−1.0, 1.0]. These values were chosen so that A consisted of individually stable parts. In ecological terms, these choices mean that the population size of each species in isolation would be stable.

Box 9.2 Generalized Lotka-Volterra Equations for *n* Species

The model in Box 9.1 can be generalized (see Taylor, 1988a, 1988b) to *n* species that potentially each interact with each other:

$$\frac{dN_i(t)}{dt} = N_i(t)\left(a_i + \sum_{j=1}^{n} \alpha_{ij} N_j(t)\right), \, i,j = 1, 2, ..., n,$$

where N_i is the population size of species *i*, a_i is the intrinsic rate of increase (or birth/death rate) of species *i*, α_{ij} is the effect of species *j* on species *i*, and *n* is the number of species.

The equilibrium population sizes, $\hat{N}_i$, are the solutions to $\frac{dN_i(t)}{dt} = 0$.

This model can also be written in matrix notation, with the community matrix *A*, consisting of the α_{ij} values, etc. The vector of equilibrium population sizes, $\hat{N} = (\hat{N}_1, \hat{N}_2, ..., \hat{N}_n)^T$, is then the solution of $A\hat{N} = -a$, where $a = (a_1, a_2, ..., a_n)^T$.

In the simplest version of this model (e.g., the equations in Box 9.1) we can constrain the self-interaction terms, α_{ii}, to be zero. In most of the models discussed in this chapter, however, no such assumption is made. For instance, in Tregonning and Roberts (1979), these terms were constrained to have magnitude one, and a stable, feasible equilibrium required species to be self-regulating, i.e., $\alpha_{ii} = -1$. The proportion of non-zero α_{ij} values with $i \neq j$ is known as the "connectance" (Gardner & Ashby, 1970; Tregonning & Roberts, 1979). It seems fair to argue that models with greater connectance are more complex, since more of the species have an effect on others.

Furthermore, this model can be scaled so that various useful analyses, such as those exploring the consequences of different parameters' values, can be carried out (Taylor, 1988a).

In explicating the work to date, May (1973) made an important general point. The argument is not about whether or not the Lotka-Volterra or any other models accurately characterize the behavior of actual biological systems. Rather, the theoreticians' results show that for simple mathematical models (including Lotka-Volterra and its more generalized form; see Box 9.2), those with few species are more stable than the corresponding more speciose models. In addition, community models with more and stronger inter-species interactions are less stable. Thus, the paradox is that we do not have a theoretical understanding of why communities that are more complex (i.e., speciose or strongly interacting) appear to be more stable.

The reaction to May's (1973) book was immediate and varied. Some of the debate centered on the details of the models; other researchers delved into ensuring the conclusions were robust to different modeling approaches (see Box 9.3). May (1973: 174) himself suggested that the way forward was to search for "devious strategies" (e.g., structuring the community into subsets or blocks of species that interact only within that block) that ensured that complexity did, in fact, allow stability. These various responses were sometimes overlapping. For instance, Roberts's (1974) emphasis on the importance of ensuring equilibrium population sizes were positive (feasibility; see Box 9.3) can be seen as all three: a detail in the model, but a check on the robustness of the conclusions, as well as a potential strategy to engender stability.

Box 9.3 The Devil in the Detail: Was May's Conclusion Right?

Roberts (1974) noted that neither the general results of Gardner and Ashby (1970) nor those of the more ecologically motivated findings in May (1972) required that the equilibrium whose stability was examined should have positive x values. Clearly, however, this restriction arises naturally when modeling population sizes. Roberts (1974) termed this positivity condition "feasibility." He then examined the generalized Lotka-Volterra equations described in Box 9.2, assuming all the birth rates (the a_i values in Box 9.2) were 1 and all the intra-specific interactions (the α_{ii} values) −1, conditions that enhance the chances of feasibility. The inter-specific interactions (the α_{ij} values for $i \neq j$) were equally likely to be either z or $-z$ (with z fixed for each system examined). Roberts's computer simulations showed that smaller z values (weaker interactions) led to more stability (as May (1972) had found in his models) but also that when only feasible equilibria were examined (which also happened more often when z was small), stability was more likely for larger systems. Roberts (1974) concluded that larger (more complex) systems exhibit greater stability.

Gilpin (1975) re-examined Roberts's model, as well as a second model in which, with equal probability, the birth rates and the intra-specific interactions were (independently) either 1 or −1. Although he confirmed Roberts's findings, they did not apply to the second model, which Gilpin claimed was more realistic (e.g., in not assuming all the species were autotrophic, $a_i > 0$). Hence, Gilpin argued, Roberts's conclusion was not robust to biologically plausible departures from his assumptions. More importantly, he pointed out that Roberts's logic did not follow. Indeed, in both models, the proportion of random models that were simultaneously feasible and stable decreased as the number of species increased.

De Angelis (1975), using a food-web model with trophic structure, found that under certain plausible conditions (e.g., stronger self-regulation at higher trophic levels), increased connectance led to greater stability. Nevertheless, increasing species number still led to a catastrophic decline in the likelihood of stability.

It is important to recognize that all of these responses continued to use an equilibrial, ahistorical framework. Although May (1973: 3) noted that the biological systems we encounter are "ones selected by a long and intricate process," he did not truly envisage using an historical approach to resolving the paradox. Rather, evolution would have generated "special sorts of complexity" and we should try to elucidate just what these conditions are, before checking that they do permit stability. Crucially, these special conditions would be found by examining the parameter values and structures of the particular models that did attain stable equilibria. The hope was that these models would show some general features (e.g., a blocked structure) that would allow some prediction of the stability of other models with (or without) these properties.

May's roadmap was followed by, for example, Tregonning and Roberts (1978), who used Monte Carlo simulations to examine the stability properties of a vast number of generalized Lotka-Volterra models with randomly generated sets of parameters (the a_i and α_{ij} values of Box 9.2). They found, for instance, that stable systems almost always consisted of self-regulating species (i.e., $\alpha_{ii} = -1$), with producers (species with $a_i = 1$) vastly outnumbering consumers ($a_i = -1$). In other words, the stable systems were structurally distinct from the overall set of models. The "unmistakably ecosystem-like" properties pertained even in a model that was seen by some as a poor reflection of biological reality.

Nevertheless, no matter what adjustments scientists made to their models, stable systems with multiple species remained very hard to find. Some other "devious strategies" were apparently going to be needed.

9.3 The Development of Ecological Complexity

To my knowledge, the first example of an historical approach to ecology's stability-complexity paradox was that of Tregonning and Roberts (1979). As before, they examined the generalized Lotka-Volterra model of Box 9.2. But, strangely (from my point of view, at least), they did not examine how multi-species communities might develop from simpler systems over time. Rather, they asked how a stable community could emerge from the successive elimination of species originally in some larger system: build down, not build up.[3]

Starting with the 50-species model, Tregonning and Roberts (1979) investigated 25 instances that differed only in the parameter values assigned at random (with some plausible constraints, such as equal numbers of consumers and producers; connectance fixed at 0.2). They first found the (unique) equilibrium of their random system (see Box 9.2), and then tested the equilibrium for feasibility (i.e., $N_i > 0$ for all $i = 1, 2, \ldots, 50$). If any population had a negative equilibrium size, the species with the most negative value was eliminated, and the appropriately reduced 49-species system again solved for its equilibrium and tested for feasibility. This process was repeated until all equilibrial population sizes were feasible, by which time they were always also stable [4]. This procedure consistently led to stable, feasible equilibria with many, many more species than previously examined "random" models (e.g., May, 1972). The 25 replicate runs culminated in "survivor systems" (as they were called by Roberts and Tregonning, 1980) with between 21 and 29 species!

Moreover, the non-random, selective nature of the elimination (i.e., removing the species with the most negative equilibrium size each iteration) was critical. If the existing species were removed with equal probability, only between two and four remained at the end.

The insight that an historical approach could produce stable multi-species ecosystems allowed Roberts and Tregonning (1980) to pursue May's agenda. They found, for

---

[3] Although Gilpin and Case (1976) had earlier used a build-down model, they were not primarily concerned with the stability-complexity paradox. Rather, their motivating questions were about the number of possible different equilibria given different starting population sizes: history as contingency, not construction. See Chapter 3.

[4] That the final feasible systems all turned out to be stable is not a given, but it seems linked to Roberts's (1974) earlier findings about feasibility and stability.

example, that the survivor systems were surprisingly robust to certain perturbations, such as the random removal of some of the survivor species. (Of course, if sufficiently disrupted, most systems, both real and model, would indeed collapse.) In other words, the theoretical survivor systems finally seemed to match the empirical ecological data as well as ecologists' intuition. At the same time, the inference had subtly changed: instead of "diversity begets stability," we had moved on to "stability permits diversity."

9.4 Construction of Ecological Complexity

The original application of a build-up model to examine the stability-complexity paradox was that of Robinson and Valentine (1979). Interestingly, these authors began their paper by discussing, in the context of ecological communities, the distinction between the number of species and the species composition. Systems resistant to changes in the latter they termed invulnerable. The concept of stability pertained to the mathematical properties of an equilibrium of an existing system and whether or not a small perturbation died out over time (see local stability in Chapter 2). This issue directly parallels the distinction between levels of genetic variation (or numbers of alleles) and genetic makeup (or allelic composition) in population genetics (see Box 8.3).

Robinson and Valentine examined yet another version of the Lotka-Volterra equations, with constraints on the constants parameterizing competition (see Box 9.4 for details). But their build-up model examined only the potential addition of a single species, not a series of potential invaders. Nevertheless, they concluded that, over time, complexity develops through the addition of new species, to produce communities that, because they become more resistant to invasion, appear stable.

> **Box 9.4 The Model of Robinson and Valentine (1979)**
>
> Robinson and Valentine took the system
>
> $$\frac{dN_i}{dt} = u_i + \sum_{j=1}^{n} a_{ij}N_j \qquad i = 1, 2, \cdots, n,$$
>
> where, as before, N_i is the population density of species i, u_i a density independent effect, such as death or immigration (for negative and positive values) and a_{ij} the interaction between species i and j. Complexity was defined as a function of both species number, n, and connectance, c, to be the number of non-zero inter-species interactions, $cn(n-1)$. To investigate "random" communities, the intraspecific interactions, a_{ii}, were sampled from the uniform distribution on $[-1.0, -0.1]$; a proportion c (the connectance) of the interspecific interactions, a_{ij} ($i \neq j$), were sampled from the uniform distribution on $[-0.7, 0.7]$, with the remaining values set to zero, and the N_i were chosen from the uniform distribution on $[100, 1000]$. The u_i were then calculated to satisfy $dN_i/dt = 0$ *for all i*, thus guaranteeing a

Box 9.4 Continued

feasible equilibrium solution. Thus, the u_i values are not random, in the sense that they are constrained to ensure each A matrix (i.e., set of a_{ij} values) generates a feasible equilibrium. If the equilibrium were also stable (ascertained by examining the eigenvalues of A), this model was retained for their analysis.

The invasion of a new species was then simulated by adding a new row and column to A (with randomly selected values) and, if possible, the new u_{n+1} was then chosen to ensure feasibility, as before. The expanded matrix was then analyzed for stability.

If this new system were stable, the invasion was successful and no existing species became extinct: the community was classed as "elastic." If not, then the dynamics of the invader were examined. A decrease implied that the community had repelled the invader and it was classified as "inelastic and invulnerable." If the invader's density increased, while that of an existing species decreased, however, the invasion caused a turnover of species, and the community was classified as "inelastic but vulnerable."

Robinson and Valentine showed that, in their model, both stability (given feasibility) and elasticity decreased with increasing complexity, as expected, but the latter more slowly. From these results they inferred that this sort of invasion process would likely lead to an increase in complexity. Moreover, invulnerability increased with complexity, and this property, not mathematical stability, was responsible for the perceived association between complexity and stability: more complex communities were simply more resistant to invasion.

Post and Pimm (1983), however, did allow the long-term development of a community, subject to successive potential invasions. In a similar manner to Robinson and Valentine (1979), they constrained the randomness of their models in ways they argued were biologically reasonable. For instance, they rejected a potential invader if this led to a trophic loop (e.g., A eats B, which eats A), and they allowed only consumers rather than producers to invade. Starting with communities of six producer species and, with a classification process like that of Robinson and Valentine (1979), they used the equations of Box 9.2 to model a succession of attempted invasions of existing communities by consumer species with random competition parameters. Their approach assumed that the time between successive invasion attempts was longer than the time for the system to equilibrate (i.e., repel the invader or reach a new equilibrium including the invader, with or without the possible extinction of existing community members). In all cases, the number of species increased rapidly, far beyond what might have been expected in a similarly sized system with all parameters chosen at random. In addition, these constructed systems were more resistant to species turnover.

9.5 Numerical Simulations Generating Ecological Complexity

Taylor (1988b) extended these results using numerical simulations. Again using a generalized Lotka-Volterra model (as in Box 9.2), he started each simulation run with a single species. He then added novel (colonizing) species one at a time to the existing system and let the system iterate to a new equilibrium. Sometimes the colonizers became established in the simulated community and sometimes one or more existing members were driven to extinction. But, over time, species diversity increased, as did system complexity, which Taylor (1988b) measured using "interactance:"

$$I = z\sqrt{C(n-1)},$$

in which z was the root mean square of the non-zero inter-species interaction values (the non-zero α_{ij} values). In short, stable complex ecosystems can easily be constructed and, *contra* May, they need not comprise weakly interacting species.

One of the debates to this point (see section 9.2 and Box 9.3) had been about the range of possible values that the various parameters could take. The constraints put on parameter values obviously affect the behavior of a model. Taylor (1988a, b) argued that there should be as few constraints as possible; that way, the values preserved in the simulated communities can then be examined for any emergent properties. To this end, he used a scaled version of the Lotka-Volterra model, sampling the novel species' ecological parameters from broader distributions than previous workers had used.

Moreover, argued Taylor (1988b, 2018), no "devious strategies" were required, since the more complex systems (i.e., those that were more speciose and connected) emerged from simulations sampling relatively unconstrained distributions of parameters. In other words, Taylor's view can be seen as the reverse of May's imposition of devious strategies: in Taylor's modeling, these strategies (i.e., properties of the stable, feasible simulations) are a byproduct of historical construction, an incidental consequence, not a pre-determined—"Mayian"—input. As he put it (Taylor, 1988b: 587), the historical approach "selects particular stable combinations of parameters even when most systems subject to the same parameter ranges are unfeasible or unstable."

One of the issues in historically informed investigations is timescale. As Taylor (2018) noted explicitly, the emergence of stable, complex model ecosystems occurred over ecological—not evolutionary—timespans. The species interactions that characterize these systems have not attained their values through the modifying actions of natural selection over extended, evolutionary timescales. Rather, a selected group of species are preserved over ecological time, and these species possess particular ecological parameter values. Indeed, the properties of the constructed ecosystems are so special that they are highly unlikely to be found by random sampling from the universe of possible values (Taylor, 1988b).

This situation parallels the evolution of fitnesses in the population-genetic models of Marks and Spencer (1991). Although the parameters governing natural selection could be said to have evolved, every bit as much as the genetic constitution of the simulated populations (see Chapter 8), they were not fined-tuned, tweaked over evolutionary time. Rather, these values changed in a stepwise fashion as the allelic constitution of the simulated populations turned over.

9.6 The Paradox Unsolved?

Taylor's modeling cannot be said to have resolved the ecological paradox, however. Eventually, in most of his computer runs, one of the species' populations expanded without limit; it "exploded" (Taylor, 1988b). Clearly, such a pathological outcome is utterly unrealistic[5]. Something critical must be missing from the modeling. Taylor pointed out that previous explorations of build-up models (e.g., Post & Pimm, 1983; Robinson & Valentine, 1979) had prevented such behavior by restricting parameter values and/or examining only feasible systems.

The most obvious route to resolution of the problem of exploding systems is clearly to restrict the universe of possible parameter values. Explosions occurred because too many positive growth rates (the a_i values of Box 9.2) and species interactions (the α_{ij} values) were selected. Constraining these distributions (perhaps by some consideration of energy conservation, which would limit the proportion of positive values) might well prevent explosions (Taylor, 1988b). Such an approach, it is worth acknowledging, is a step towards May's devious strategies, however!

9.7 Current Questions

In the 35 years since Taylor's work, there has been significant further research under the rubric of "community assembly." This book is not the place to detail that history, but the interested reader can consult reviews in Warren (1994), Brännström et al. (2012) or Shinohara et al. (2023). One aspect noteworthy from a history-of-science perspective is how the historical approach is now thoroughly embedded into mainstream ecological thinking, Taylor's (2018) concerns notwithstanding. (A recent quick search on Web-of-Science using the term "community assembly" returned > 6500 publications.)

Some questions that appeared to have been settled have repaid a revisit. Stone (2018), for instance, examined models of density-dependent communities underpinned by random interaction matrices, and revealed an intriguing link between feasibility (i.e., having populations with positive equilibrium sizes; see Box 9.3) and stability. As Roberts (1974) found, almost all feasible systems are stable, but constructing feasible systems is extremely difficult. Consequently, complex yet stable systems may be even rarer in parameter space than May predicted.

Most exciting, though, are insights that would never have been possible without taking history seriously. For example, Maynard et al. (2018) have shown that a

---

[5] Such behavior was not possible in the population-genetic models of Chapter 8, since the equations governing the allele frequencies map the simplex $\Delta = \{(p_1, p_2, \cdots, p_n) : 0 \leq p_i \leq 1 \ \forall i \ \& \ \sum_{i=1}^{n} p_i = 1\}$ onto itself.

number of patterns frequently observed in stable ecological networks are not selected directly for their stability-enhancing properties during community assembly, as has often been surmised. These workers used two different scenarios for community assembly. The first, "immigration" of species with no relationship to resident species, modeled the potential invaders with their interaction parameter values chosen independently, as is the case in the models discussed above. The second was dubbed "radiation" and envisaged a randomly chosen resident species to be parental to a daughter species, the potential invader, with interactions being small perturbations of those of the parent. The two scenarios resulted in very different properties of the interaction matrices of the final (stable) communities; these differences were reflected in significant differences in the structures of the communities (see Box 9.5 for details). Crucially, by comparing the outcomes with random (null) expectations, Maynard et al. (2018) showed that these different structures are the incidental consequences— "spandrels" *sensu* Gould and Lewontin (1979)—of the way in which invading species enter the systems, rather than the direct result of selection for stability.

Box 9.5 Eigenvalues and the Models of Maynard et al. (2018)

In almost all of the mathematical models in this and other chapters, the parameters of the model can be depicted in matrix form. Moreover, this matrix is almost always square: it has the same number of rows as columns. In this chapter, the ecological interaction parameters (e.g., the α_{ij} values of Box 9.2) can be placed in a square matrix, because each species potentially interacts with itself and every other species. The selection coefficients of the models of Chapter 8 also fit into a square matrix, although this matrix is symmetrical since reciprocal heterozygotes (A_iA_j and A_jA_i) are genetically identical.

The eigenvalues of a square matrix, M, are arguably its most important properties. They determine the long-term behavior of the variables in the model parametrized by M. As we saw in Box 2.3, eigenvalues can be used to determine the local stability of an equilibrium. Effectively this determination is possible because the eigenvalues tell us whether deviations from the equilibrium (in the values of the variables) will shrink or grow. For details about the calculation and interpretation of eigenvalues, see Roughgarden (1979), Edelstein-Keshet (1988) or Otto and Day (2007).

In the models examined by Maynard et al. (2018), stability necessitated that the eigenvalues of the interaction matrix be negative. The eigenvalues from the "immigration" scenario exhibited a distribution of negative values, which was indistinguishable from that for random matrices with the same mean and variance of their elements. Those from the "radiation" scenario showed a very different distribution, although, of course, of only negative values. The appropriate random distribution for comparison, however, included positive values, and this comparison might be thought (erroneously) to imply that there has been selection for stability under radiation. But the difference clearly arises as a byproduct (a "spandrel") of the two different assembly scenarios, not as a direct result of selection for stability.

Network diagrams showing the strongest 10% of inter-species interactions as edges between nodes representing species (see Fig. 9.2) revealed that the immigration scenario led to graphs that appeared random, whereas radiation produced highly structured systems

Box 9.5 *Continued*

consisting of clusters of species that interacted mostly within their own cluster. Again, this property is clearly a spandrel of the assembly process; it is not directly selected.

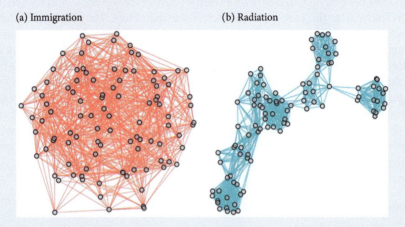

Fig. 9.2 Interactions (edges) between species (nodes) in two scenarios of community assembly, (a) "Immigration" and (b) "Radiation," in the models of Maynard et al. (2018). The obvious differences are a spandrel of the assembly process, not a directly selected property. See text for details.

Reproduced from Maynard et al. (2018) with permission.

Similar conclusions were reached by Sarevia et al. (2022), who compared empirical food webs with those assembled by sampling species (and their interaction parameters) from the empirical pool of potential species (the "metaweb"). They found almost no differences and hence argued that, rather than being directly selected, the structure and, especially, the stability observed in food webs was merely inherited from the metaweb.

These findings have great import for the intellectual descendants of May's roadmap: we cannot simply look at the properties of interaction matrices that appear to confer stability and assume these properties are the product of a selection. In many cases they may simply be spandrels of the assembly process.

9.8 Priority Effects

Much of the above discussion may remind readers of "priority effects" (a term apparently first used by Shulman et al., 1983), the consequences of the order or timing of the arrival of potentially invading species in ecological succession. Nevertheless, taking as a premise that systems had more than one stable equilibrium, much research on priority effects has focused more on which species successfully invade an existing community and hence its final composition rather than (as in the work discussed in this chapter) the issue of stability versus complexity.

For example, Kilsdonk and De Meester (2021) were interested in the genetic makeup of a population assembled over time from a pool of genetically variable immigrants. Their modeling showed that priority effects were critical: often such populations were dominated by the descendants of early colonists, which were able to adapt to local conditions and exclude subsequent immigrants. The order and rate at which immigrants arrived determined their evolutionary success. Thus, the differences among local populations may be explained by priority effects as much as traditional founder effects, which imply a much lower dispersal rate.

Similarly, Weidlich et al. (2021) recently argued that priority effects could be used as an ecological-restoration tool, promoting the establishment of the species being reintroduced and hampering the chances of undesired taxa invading. They introduced priority effects as an aspect of contingency, and this classification fits well with mine. Similarly, in their review of community assembly, Shinohara et al. (2023) emphasized the importance of contingency in generating the variation in membership of assembled systems.

These questions, it seems to me, are about which of a possible range of different equilibria are attained, and hence have more in common with the foraging-fish example of Chapters 1 and 3 than they do with those about the construction of complexity and its association with stability. Hence, I would regard priority effects as examples of historical contingency.

9.9 Ecological Construction in Perspective

As with theoretical evolutionary genetics, ecological theory has greatly benefited from the recognition that more complex mathematical systems can be built up over time. This construction of complexity arises because of the trial-and-error mechanism inherent in the arrival of colonizing species and the subsequent population dynamics of the set of species present. The latter process eliminates species that do not permit stability. The constructionist approach in ecology has at least partially solved the paradox of complexity not implying stability. Moreover, it has enabled the asking of deeper questions about the structure of stable mathematical models, questions that an equilibrial ahistorical approach would never have raised.

References

Brännström Å., Johansson J., Loeuille N., Kristensen N., Troost T. A., Hille Ris Lambers R., Dieckmann U. 2012. Modelling the ecology and evolution of communities: A review of past achievements, current efforts, and future promises. *Evolutionary Ecology Research* 14:601–625.

Edelstein-Keshet L. 1988. *Mathematical Models in Biology*. Boston, Mass.: McGraw-Hill.
Elton C. S. 1958. *The Ecology of Invasions by Animals and Plants*. London: Methuen.
Gardner M. R., Ashby W. R. 1970. Connectance of large dynamic (cybernetic) systems: Critical values for stability. *Nature* 228:784.
Gilpin M. E. 1975. Stability of feasible predator-prey systems. *Nature* 254:137–139.
Gilpin M., Case T. 1976. Multiple domains of attraction in competition communities. *Nature* 261:40–42.
Gould S. J., Lewontin R. C. 1979. The spandrels of San Marco and the Panglossian paradigm: A critique of the adaptationist programme. *Proceedings of the Royal Society of London B* 205:581–598.
Hutchinson G. E. 1959. Homage to Santa Rosalia or why are there so many kinds of animals? *American Naturalist* 93:145–159.
Kilsdonk L. J., De Meester L. 2021. Transient eco-evolutionary dynamics and the window of opportunity for establishment of immigrants. *American Naturalist* 198:E95–E110.
MacArthur R. H. 1955. Fluctuations of animal populations, and a measure of community stability. *Ecology* 35:533–536.
Marks R. W., Spencer H. G. 1991. The maintenance of single-locus polymorphism. II. The evolution of fitnesses and allele frequencies. *American Naturalist* 138:1354–1371.
Marshall I. 2017. Kurt Vonnegut's "Homage to Santa Rosalia": The "Patroness of Evolutionary Studies" and *Galapagos*. *Evolutionary Studies in Imaginative Culture* 1:137–147.
May R. M. 1972. Will a large complex system be stable? *Nature* 238:413–414.
May R. M. 1973. *Stability and Complexity in Model Ecosystems*. Princeton, NJ: Princeton University Press.
Maynard D. S., Serván C. A., Allesina S. 2018. Network spandrels reflect ecological assembly. *Ecology Letters* 21:324–334.
Otto S. P., Day T. 2007. *A Biologist's Guide to Mathematical Modeling in Ecology and Evolution*. Princeton, NJ: Princeton University Press.
Post W. M., Pimm S. L. 1983. Community assembly and food web stability. *Mathematical Biosciences* 64:169–192.
Roberts A. 1974. The stability of a feasible random ecosystem. *Nature* 251:607–608.
Roberts A., Tregonning K. 1980. The robustness of natural systems. *Nature* 288:265–266.
Robinson J. V., Valentine W. D. 1979. The concepts of elasticity, invulnerability and invadability. *Journal of Theoretical Biology* 81:91–104.
Roughgarden J. 1979. *Theory of Population Genetics and Evolutionary Ecology: An Introduction*. New York, NY: Macmillan.
Saravia L. A., Marina T. I., Kristensen N. P., De Troch M., Momo F. R. 2022. Ecological network assembly: How the regional metaweb influences local food webs. *Journal of Animal Ecology* 91:630–642.
Shinohara N., Nakadai R., Suzuki Y., Terui A. 2023. Spatiotemporal dimensions of community assembly. *Population Ecology* 65:5–16.
Shulman M. J., Ogden J. C., Ebersole J. P., McFarland W. N., Miller S. L., Wolf N. G. 1983. Priority effects in the recruitment of juvenile coral reef fishes. *Ecology* 64:1508–1513.
Stachowicz J. J., Whitlatch R. B., Osman R. W. 1999. Species diversity and invasion resistance in a marine ecosystem. *Science* 286:1577–1579.
Stone L. 2018. The feasibility and stability of large complex biological networks: A random matrix approach. *Scientific Reports* 8:8246.

Taylor P. J. 1988a. Consistent scaling and parameter choice for linear and generalized Lotka-Volterra models used in community ecology. *Journal of Theoretical Biology* 135:543–568.

Taylor P. J. 1988b. The construction and turnover of complex community models having generalized Lotka-Volterra dynamics. *Journal of Theoretical Biology* 135:569–588.

Taylor P. J. 1989. Developmental versus morphological approaches to modelling ecological complexity. *Oikos* 55:434–436.

Taylor P. J. 2005. *Unruly Complexity: Ecology, Interpretation, Engagement.* Chicago: University of Chicago Press.

Taylor P. J. 2018. From complexity to construction to intersecting processes: Puzzles for theoretical and social inquiry. *Ecological Complexity* 35:76–80.

Tregonning K., Roberts A. P. 1978. Ecosystem-like behaviour of a random interaction model. I. *Bulletin of Mathematical Biology* 40:513–524.

Tregonning K., Roberts A. P. 1979. Complex systems which evolve towards homeostasis. *Nature* 281:563–564.

Warren P. H. 1994. Making connections in food webs. *Trends in Ecology & Evolution* 9:136–141.

Weidlich E. W. A., Nelson C. R., Maron J. L., Callaway R. M., Delory B. M., Temperton V. M. 2021. Priority effects and ecological restoration. *Restoration Ecology* 29:e13317.

10

Concluding Remarks

10.1 Why Bother with Flavors?

In the preceding chapters, I have described a number of historical flavors as I see them and, for each of these flavors, illustrated, with a number of examples, how an approach that takes account of the importance of history has resulted in a better understanding of the relevant eco-evolutionary system. My hope is that, by choosing a variety of exemplars, different cases resonate with different readers, who are then stimulated to think about their favorite study system in an historical manner, in addition to the standard ahistorical, equilibrium-centered view. Naturally, some flavors will be more important than others in any particular study system and, indeed, overall in ecology and evolution. For instance, chaos has not played a major role in any of my own research.

Nevertheless, I am sure that few of these cases were informed by explicit awareness of the importance of history. Some clear exceptions, in which history was overtly invoked, are several of the models of Chapters 3 (the contingency in the model of competitive foraging fish and the different outcomes in competitive communities), 8 (the construction of single-locus polymorphisms) and 9 (the construction of ecological complexity in model ecosystems). Furthermore, even fewer cases have differentiated among the different flavors I have identified. Perhaps Lewontin's (1966) characterization of capriciousness as being less certain than chance is the notable early exception here.

So, then, what is the point of distinguishing among the flavors of history? The answer, as I suggested in Chapter 1, is that the explicit consideration of different flavors is of practical benefit in understanding the role of ecological and evolutionary history in our study systems. Lewontin's example of capriciousness, the population-genetic consequences of reversing the order of random selection coefficients (detailed in Chapter 6), led him to describe the "principle of historicity" and the critical distinction between stochasticity and capriciousness. A failure to recognize these two flavors would mean that we did not really understand his system, and we would have been satisfied with some of the various plausible, yet erroneous, explanations (e.g., genetic drift, differing selection pressures) for the contrasting outcomes of the opposite environmental sequences.

10.1.1 Contingency versus Construction

Different flavors also motivate different questions. Although the flavors of contingency and construction can both be present in models of community assembly (see Chapter 9), they direct our attention to different aspects of the process. In our ecological models, contingency prompts questions about which species are present in our systems, whether different equilibria are possible and, if the system is at equilibrium, why we have arrived at that equilibrium and not another. Construction, however, leads us to ask about the properties of our system: How many species are present? Is it stable (both internally and to invasion by new species)? And, if so, what properties does it possess that confer that stability?

The constructionist population-genetic models discussed in Chapter 8 were designed to address the Dobzhanskian problematic of why we observe so much genetic variation in natural populations. In particular, these models examined the possibility, favored by Dobzhansky, that selection was responsible for maintaining numerous alleles at a locus. Thus, history as construction focused on the numbers of alleles, and the allele-frequency distributions and sorts of fitness sets, that underlaid the constructed polymorphisms. In their models of frequency-dependent selection, for example, Trotter and Spencer (2008) found that their simulated populations possessed more alleles after 10,000 generations than the comparable constant-viability models of Marks and Spencer (1991). The latter workers reported significant correlations between heterozygote fitnesses and the average of the two viabilities of the associated homozygotes at the end of their simulations.

But it is also possible to ask contingency-inspired questions about the results of such simulations. Clearly, which mutants successfully invade the simulated polymorphisms is contingent on those that are already there. For example, we might be interested in the relatedness and genealogy of alleles in the system. In this case, we might examine a model in which each mutant allele is derived from an existing, parental allele in the population (e.g., Spencer & Marks, 1992; Trotter & Spencer, 2013). Pheno/genotypes involving such mutants likely have fitnesses that are some small perturbation of those involving the parental allele. So, for example, the fitness of $A_{n+1}A_j$ would be close to (but on average a little smaller than) that of A_pA_j, where A_p was the parental allele that mutated into A_{n+1}. We can then ask whether the simulated polymorphisms are dominated by closely or distantly related alleles, those with similar or very different fitnesses.

It is natural, then, to compare these theoretical phylogenies with the results from real populations in which we have sequenced the different alleles at a single locus. Assuming a simple, parsimonious model of molecular evolution that allows evolutionary relationships and their underlying changes to be deduced, we can estimate the allelic genealogy (Leigh & Bryant, 2015). Indeed, variation in mitochondrial genes is often portrayed this way, in so-called "haplotype networks." The estimation of such trees for nuclear genes is more complicated because of the possibility of intragenic recombination (which, critically, is absent from the existing theoretical models of construction discussed above).

These questions are paralleled to a remarkable extent by some older work on the chromosomes of the sister-species pair, *Drosophila pseudoobscura* and *D. persimilis*. As part of an unmatched series of papers, Dobzhansky and his co-workers studied, among other things, the naturally occurring inversions of chromosome 3 found in natural populations of these cryptic fruit-fly species. Staining of polytene chromosomes from the flies' salivary glands allows these chromosomal arrangements to be visualized, revealing unique banding patterns, which, together with any overlaps, allows inferences about their genealogical relationships (Fig. 10.1). Indeed, it is even possible to infer the direction of the changes (Powell, 1992, 1997).

What I find remarkable about Fig. 10.1 is that it shows but a single "hole." Only one unobserved inversion, dubbed "Hypothetical," is required to connect all 50-odd naturally occurring arrangements by a series of single inversion events. This pattern implies a number of important properties of these inversions, notably that almost all of them (and certainly those from which other inversions have arisen) are likely to be selectively advantageous in (at least) some circumstances (Powell, 1992, 1997). Note that this method of portraying chromosomal variation points us directly toward the selective hypothesis. In fact, both cage experiments and observations in the wild (e.g., Dobzhansky, 1948a, 1948b) have shown that selection does indeed have a hand in molding some of this character.

The perspicacious reader will have noticed that, in asking what we can deduce from an allelic or chromosomal genealogy, we are dealing with yet another flavor of history, history as template. In terms of this book's theme, then, we should note the more general point, namely that the consideration of one flavor of history (here, originally construction) can generate questions that pertain to other flavors (in this case, first contingency and subsequently template).

10.2 Multiple Flavors

The above linking of several flavors of history naturally leads to a further reason for bothering about flavors: it can alert us to multiple aspects of history in one single study system. For example, Trotter and Spencer (2008, 2013) sought to quantify the ability of frequency-dependent selection to maintain allelic variation. They explicitly followed the constructionist approach outlined in Chapter 8, bombarding their simulated populations every generation with a novel mutant, before selection took place. They then examined the resultant populations after 10,000 generations, characterizing the number of alleles and the distributions of allele frequencies and the fitness parameters. So far, this work is standard history as construction.

Trotter and Spencer (2008, 2013) were aware, however, that these simulations may not have been at an equilibrium: they may have been experiencing history as turnover. To confirm this possibility, Trotter and Spencer (2008, 2013) took the populations at Generation 10,000 and let them move to a selective equilibrium, by continuing the

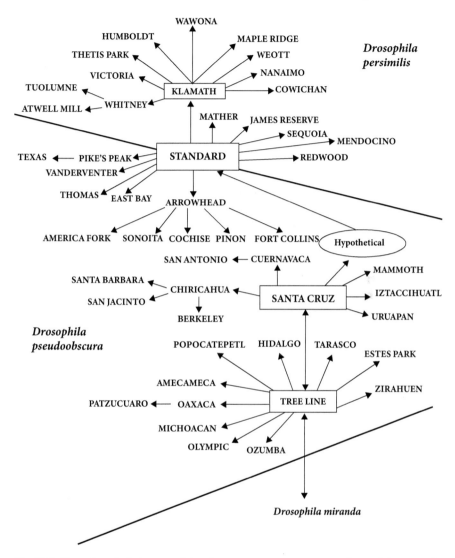

Fig. 10.1 The inferred phylogeny of the chromosome-3 inversions in *Drosophila persimilis* and *D. pseudoobscura*. Each of the names represents a particular chromosomal arrangement, with each arrow indicating the origin of a new arrangement via a single inversion event. Note that "Standard" is found in both species. There is just one "hole" in this phylogeny: "Hypothetical" bridges the two-inversion change between "Santa Cruz" and "Standard."

Adapted from Powell (1997) with permission.

iterations in the absence of further mutation. And, indeed, they found a substantial drop in the mean numbers of alleles present as some alleles were eliminated (see Fig. 10.2). In short, the genetic constitution of the populations in the full simulations

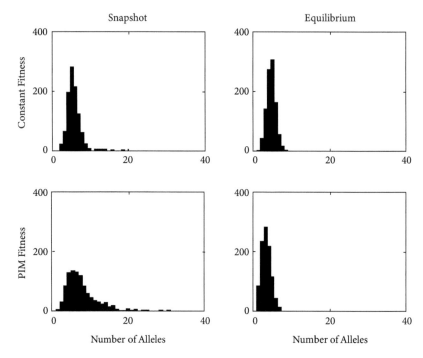

Fig. 10.2 The distribution of the number of alleles produced by 1000 replicate runs of (top) the constant-viability model of Spencer and Marks (1988) and (bottom) the frequency-dependent pairwise-interaction model (PIM) of Trotter and Spencer (2008). The left-hand panels show these distributions at Generation 10,000 ("snapshot"); the right-hand panels show the distributions after each simulation is allowed to run to equilibrium in the absence of novel mutation. Note that the distributions are very similar for the models of constant-viability selection, with the greatest difference being the elimination of outcomes with larger numbers of alleles (> 10) in the equilibrium distribution. By contrast, in the models of frequency-dependent selection, there is a significant shift to the left in the equilibrium distribution compared to that of the snapshot.

(i.e., including mutation, up to Generation 10,000) remained dynamic. Some alleles were on the road to extinction at the same time as others were increasing in frequency; even as polymorphism was being constructed, the elements of that polymorphism were being replaced. This insight shows how even simple models of selection may lead to allelic turnover in a similar manner as in Kimura's neutral model. And, indeed, the decline in fitness differences among different extant genotypes as the selective simulations progress suggests a further parallel, namely that polymorphisms constructed under a selective regime may evolve to be almost selectively neutral.

So here we are witnessing two flavors, construction and turnover, and it does not take much imagination to see how contingency and template might also be relevant. Each of these flavors prompts different questions, and it is through the totality of the corresponding answers that we can fully appreciate what is happening in our study system. And even though it may be possible to arrive at many of these conclusions without the explicit use of historical flavors, it seems to me that they provide a useful framework to get us to think systematically and critically about how history may be important in biology.

A second example, from ecological theory, is provided by Becker et al. (2022), who examined a model of community assembly incorporating stochastic population dynamics, thus blending construction and chance. Among other things, these authors showed that stochasticity in the form of demographic noise was essential for the maintenance of both specialists and generalists. Moreover, this diversity arose from the ongoing turnover of species, a third of my historical flavors. In addition, the models made predictions consistent with the intermediate disturbance hypothesis (e.g., that species diversity is maximized at intermediate stages of assembly). Once again, an historical perspective reveals its explanatory power.

10.3 The Historical Dimension

At a more general level, the breadth of examples discussed in the previous chapters shows the importance of taking an historical perspective in ecological and evolutionary studies. Although we have the benefit of hindsight, it seems to me that the focus on equilibria inherent in much mathematical ecology and evolution in the twentieth century should have been sufficient to get theorists (and, indeed, empiricists) to realize that history was important in scientific understanding of the biological realm. After all, the existence of an equilibrium implies that there may have been a past system. But, as Kingsland (1985) explains, the focus on equilibrium (in theoretical ecology, at least) was deliberately ahistorical—anti-historical, one could say—designed to allow research to ignore history and its supposedly unscientific contingency.

The legacy of failing to account properly for the effects of history is with us even today. Amongst evolutionary biologists, perhaps those interested in genome evolution have been the most aware of the importance of history. But too many evolutionary geneticists, in my view, still seem unaware of how taking history seriously would improve their science; a little ironic considering the inherently historical nature of organic evolution. For instance, why do population geneticists continue to cite the ahistorical result of Lewontin et al. (1978)—about the small size of the parameter space affording stable feasible equilibria in simple models of viability selection—as evidence for the improbability of the selective maintenance of genetic variation, when that line of reasoning is logically flawed?! This finding is incredibly important—don't get me wrong—but it does not (and cannot) inform us about the likelihood of our historical, evolutionary models seeking out these special regions. Moreover, there are far better, logically defensible arguments against the efficacy of selection maintaining genetic variability.

Ecologists have probably been faster than evolutionists to appreciate the relevance of the historical dimension. The concept of community assembly and associated ideas, for example, has blossomed to the point where it is now a whole subfield of theoretical ecology, asking detailed and sophisticated questions. Perhaps a more general awareness of the importance of the past among ecologists (see Beller et al. 2017) is responsible. Nevertheless, it still seems to me that much ecological research would benefit from a more explicit incorporation of an historical perspective.

10.4 A Roadmap for Researchers

So, what should researchers actually do? In short, my view is that we should always ask whether our study system has any sort of historical dimension. In ecology and evolution, that is very likely to be true, and so then the question becomes how might history have been important. It is at this point that the use of flavors is most likely to help. Once we have resolved this matter, we can then ask how we can incorporate such an understanding into our investigations. For an illustrative example from my own recent work, see Box 10.1.

Box 10.1 Constructing Single-locus Polymorphism in a Deteriorating Environment

The constructionist population-genetic models discussed in Chapter 8 largely assume that the environment—and hence also the selection parameters—remains constant over time. Even the models exploring frequency-dependent selection assume constant interactions between pheno/genotypes. Consequently, populations in these latter models with the same frequencies of particular pheno/genotypes will have the same fitnesses, even if separated in time for thousands of generations. Ironically, all these models ignore a pervasive aspect of biological history, namely environmental deterioration and the concomitant decrease in fitnesses of existing types in a population over time. Such historical changes can arise from either abiotic factors (as physical resources are depleted or climate changes) or biotic factors (as competitors, predators, prey and parasites adapt to the effects of the species or population of interest). Recognition of temporal non-constancy in evolution inspired Van Valen's (1973) Red Queen Hypothesis, in which a species had to continually adapt simply to maintain its current position.

Spencer and Walter (2024) have recently added this form of history as turnover to the constructionist modeling of Chapter 8 (section 8.4). In addition to the introduction of a new mutation in each generation of their simulations, the fitnesses of existing pheno/genotypes were decremented slightly, by multiplying each by a constant factor d (with $d \in \{0.995, 0.999, 0.9999\}$). The original Spencer and Marks (1988) models had an implicit $d = 1$. As in previous work, 10^4 replicate runs (differing only in the pseudo-random number seed) were performed. At Generation 10,000 the resulting model populations were censused and the number of extant alleles (n) and common alleles (n_c, those with frequencies > 0.1) were recorded. Previous constructionist models (e.g., Fig. 10.3a) rarely generated monomorphisms or large numbers of alleles (many of which might be uncommon, of course). We can see from Fig. 10.3b, however, that both these outcomes were observed even when the fitness deterioration was low (0.01% per generation).

Box 10.1 *Continued*

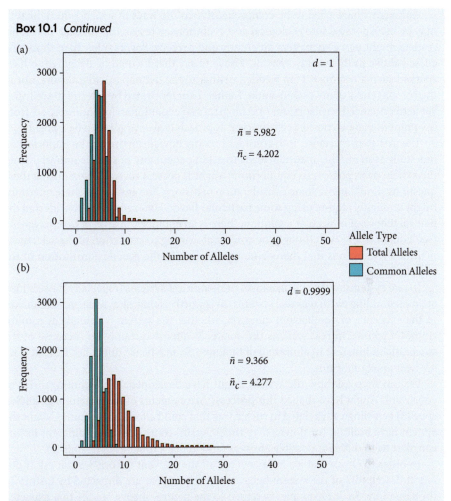

Fig. 10.3 Bar charts showing the distribution of the mean number of alleles and common alleles in the constructionist simulations of Spencer and Walter (2024), without environmental deterioration (a) and with environmental deterioration (b).

After Spencer and Walter (2024).

Incorporating this neglected historical flavor (turnover) into models that already had a constructionist flavor significantly improved the models' match with real data, which show a wide range of allele numbers. Pleasingly, too, the outcomes were reasonably robust to the addition of a third flavor, stochasticity in the form of genetic drift, as well as more sophisticated fitness structures, such as generalized dominance (see Box 8.4).

It is noteworthy that there is no obvious ahistorical way to carry out this modeling: like the neutral model of molecular evolution, there is no equilibrium to find, as the alleles constantly turn over.

Although I have tried to be comprehensive in the ways in which I think history may be incorporated into ecological and evolutionary research, I concede that some readers might perceive matters differently and perceive flavors other than those discussed above in their study systems. I have to say that I would be delighted to have sparked such a response. I am wedded neither to the use of particular flavors, nor to the completeness of my classification. Rather, I am motivated by a desire to see fuller, yet better nuanced, explanations of ecological and evolutionary phenomena, so, from my point of view, a considered historical approach is always going to be a positive.

The first step, therefore, is to consider broadly how history might be important in our study system. This naturally leads to asking about any possible equilibria. Does the system even possess an equilibrium? Might it possess several? Even raising these questions should invite some consideration of history. The answers to these questions might implicitly require using some particular flavor. For example, in considering the neutral theory of molecular evolution, there is an equilibrium level of heterozygosity (see Equation 7.1) even though the continual drifting to extinction of the selectively neutral alleles means that there is no equilibrium in the genetic constitution of the population.

In real systems or specific instances of theoretical systems (especially those aimed at understanding particular real systems), it is worth asking, also, whatever the answer to the existence of equilibrium question, whether the system is actually at equilibrium. Of course, in real systems the answer is almost certainly no (because of the stochasticity inherent in almost all of biology). So maybe we might ask instead if we are near an equilibrium.

We can then ask how the system might have developed. What contingencies or constraints might have shaped this process? Are elements of the system (e.g., alleles, species) genealogically related in such a way that could affect the outcome? Might the system have built up (or down) over time? Again, consideration of different flavors can alert us to different possible answers (and hence a fuller understanding).

In order to check that we have covered all the relevant flavors, we can ask if different descriptors of the system behave similarly and/or are impacted by history in the same way. So, for example, to return to the neutral theory, we see that expected heterozygosity, which has an equilibrium (albeit one to which the system is rarely close), gives a different historical perspective than the allelic makeup, which has no equilibrium and is constantly turning over.

In general, the way forward is to approach our problem asking as many different sorts of historically informed questions as possible. Failing to do so means that we may well arrive at completely the wrong conclusion. Of course, I am not the first to make this point. Almost 50 years ago, Gilpin and Case (1976) pointed out that even the most basic ecological questions about the presence or absence of a species needed to consider at least two historical aspects (what I call the flavors of turnover and contingency):

> Ecologists typically explain the absence of a species in an area by searching for ultimate climatic or physical factors that limit its distribution or favour its competitors. The possibility

that the system is not in equilibrium—that colonisation has not occurred—is not normally considered. Even less appreciated is that there may be a number of alternative stable equilibrium communities which may develop in a given area and that the final outcome may depend solely or partially on historical factors such as the sequence and numbers in which each species colonises.

Maybe theoretical ecologists seem to be more aware of the importance of history because of efforts such as this paper.

In short, using an explicitly historical research approach that distinguishes among flavors means that we are likely to: (i) understand our study systems more deeply and accurately; (ii) recognize that a number of different historical processes may be acting in our system; and (iii) ask a wider range of questions, most importantly questions that do not obviously arise under one particular flavor.

References

Becker L., Blüthgen N., Drossel B. 2022. Stochasticity leads to coexistence of generalists and specialists in assembling mutualistic communities. *American Naturalist* 200:303–315.

Beller E., McClenachan L., Trant A., Sanderson E. W., Rhemtulla J., Guerrini A., Grossinger R., Higgs E. 2017. Toward principles of historical ecology. *American Journal of Botany* 104:645–648.

Dobzhansky T. 1948a. Genetics of natural populations. XVI. Altitudinal and seasonal changes produced by natural selection in certain populations of *Drosophila pseudoobscura* and *Drosophila persimilis*. *Genetics* 33:158–176.

Dobzhansky T. 1948b. Genetics of natural populations. XVIII. Experiments on chromosomes of *Drosophila pseudoobscura* from different geographical regions. *Genetics* 33:588–602.

Gilpin M., Case T. 1976. Multiple domains of attraction in competition communities. *Nature* 261:40–42.

Kingsland S. E. 1985. *Modeling Nature: Episodes in the History of Population Ecology*. Chicago: University of Chicago Press.

Leigh J. W., Bryant D. 2015. PopArt: Full-feature software for haplotype network construction. *Methods in Ecology & Evolution* 6:1110–1116.

Lewontin R. C. 1966. Is nature probable or capricious? *BioScience* 16:25–27.

Lewontin R. C., Ginzburg L. R., Tuljapurkur S. D. 1978. Heterosis as an explanation for large amounts of polymorphism. *Genetics* 88:149–170.

Marks R. W., Spencer H. G. 1991. The maintenance of single-locus polymorphism. II. The evolution of fitnesses and allele frequencies. *American Naturalist* 138:1354–1371.

Powell J. R. 1992. Inversion polymorphisms in *Drosophila pseudoobscura* and *Drosophila persimilis*. Pages 73-126 in *Drosophila Inversion Polymorphism*, edited by C. B. Krimbas and J. R. Powell. Boca Raton, Florida: CRC Press.

Powell J. R. 1997. *Progress and Prospects in Evolutionary Biology: The Drosophila Model*. New York, NY: Oxford University Press.

Spencer H. G., Walter C. B. 2024. Polymorphism and the Red Queen: The selective maintenance of allelic variation in a deteriorating environment. *G3: Genes, Genomes, Genetics* 14:jkae107.

Spencer H. G., Marks R. W. 1992. The maintenance of single-locus polymorphism. IV. Models with mutation from existing alleles. *Genetics* 130:211–221.

Trotter M. V., Spencer H. G. 2008. The generation and maintenance of genetic variation by frequency-dependent selection: Constructing polymorphisms under the pairwise interaction model. *Genetics* 180:1547–1557.

Trotter M. V., Spencer H. G. 2013. Models of frequency-dependent selection with mutation from parental alleles. *Genetics* 195:231–242.

Van Valen L. 1973. A new evolutionary law. *Evolutionary Theory* 1:1–30.

Index

For the benefit of digital users, indexed terms that span two pages (e.g., 52–53) may, on occasion, appear on only one of those pages.

Tables, figures, and boxes are indicated by an italic *t*, *f*, and *b*.

A

adaptation 35, 48–51, 50*f*, 54
adaptive landscape 35–37, 36*f*
adaptive radiation. *See* adaptation
aphids 45–46, 45*f*, 46
approach, flavor of 10, 12*t*, 81–85, 88–89, 92
assembly rules 41, 108, 125, 129, 130

B

balance of nature 1, 8
Beatty, John 8, 95
Begum, Mahmuda 49–51
behaviour, evolution of 54–56, 64
bifurcation 22*b*, 41, 44
bottlenecks, population. *See also* genetic drift. 26, 47–48, 89
Buchnera 42–43, 45–46, 45*f*
buntings 30–32
Buri, Peter 69–71, 70*f*
bw[75] mutation in *Drosophila* 69–71, 70*f*

C

capriciousness, flavor of 9, 11, 12*t*, 18*b*, 67–68, 75–79, 88, 124
Case, Ted 37–38, 38*f*, 113*b*, 114 n. 3, 132
Cauchy distribution 75–76, 75*f*, 76*f*
chance, flavor of 9, 12*t*, 18*b*, 67–75, 79, 85, 88–89, 92, 124, 129, 130*b*
chaos, flavor of 9, 11, 12*t*, 18*b*, 67–68, 72–75, 79
cheats 26
Chevra Dor Yeshorim 84
Chong, Rebecca 45–46, 45*f*
cirl bunting 30–31
classical/balance controversy 8, 95–96, 99–100
Clayton, Dale 9, 57, 58*f*
coevolution of hosts and parasites 9, 12*t*, 56–57
competitive feeding fish. *See* foraging fish
complexity and stability, ecological 94, 108–109, 111, 112–114, 114 n. 1, 115*b*, 117, 120–121, 124
connectance 111–112*b*, 113*b*, 114–115*b*
constraint, flavor of 9, 12*t*, 41, 67, 69–71
construction, flavor of 6, 10–11, 12*t*, 35, 94, 108, 124, 125–126, 128–129
contingency, flavor of 1, 4*f*, 6, 8, 9, 12*t*, 29, 41, 48, 51, 67, 69–71, 110*f*, 114 n. 1, 121, 124, 125–126, 128–129, 132
co-operators 26
cormorant 54–55, 55*f*, 56, 64
cycling 15–18, 17*f*, 22*b*, 26, 73–74, 74*f*, 74, 86, 109*b*

D

deleterious recessive, model of 24–25, 25*f*, 71–72, 72*f*, 81–82, 84, 89
  with genetic drift 72, 72*f*, 85, 86*f*, 92
  with mutation 10, 82–83*b*, 84–85, 86*f*
Demodex folliculorum 47–48, 47*f*
density-dependence 8, 26, 73, 118
deterministic models, definition of 24
developmental canalization 41, 42*f*, 51
developmental noise 68–69
Diamond, Jared 41
disease epidemics, infectious 9, 12*t*, 67–68, 74–75, 79,
diversity, biological, maintenance of 94–95
Dobzhansky, Theodosius 8, 94, 95–96, 125–126
drift. *See* genetic drift
Drosophila 1–2, 69–71, 70*f*, 89, 96, 126, 127*f*
Dunham, Arthur 54, 63
Durvillaea antarctica 61, 61*f*, 63

E

ecological complexity 12*t*, 108
endosymbionts 42–43, 45–46, 57
equilibria, concept of 1, 6–7, 15
   how to find 18*b*, 82*b*
   importance of 102*b*
   multiple 25–26, 38, 38*f*
   stability of. *See* stability
exponential growth, model of 23

F

Fahrenholz's Rule 56–57
feasibility 98, 112*b*, 112–113*b*, 114, 114 n. 4, 115*b*, 117–118
fish, foraging. *See* foraging fish
fitness space 97–99, 98*f*, 99*b*, 102, 105
flavors, historical. *See also individual flavors*.
   general definition of 2, 8–11, 12*t*
   multiple 71, 72*f*, 72, 79, 85, 92, 126–129
   uses and advantages of 11, 124–126, 130–133
food-web models C 9
foraging fish 3–6, 4*f*, 9, 12*t*, 14, 25–26, 29, 32–35*b*, 124
foraging theory, optimal. *See* optimal foraging theory.
Fretwell, Stephen 3

G

gain-of-function. *See* adaptation
generalized dominance 104*b*, 104, 105*f*
generalizing 1, 29, 39
genetic drift,
   as an example of history as chance 9, 12*t*, 67, 69–71, 70*f*, 79
   in constructionist models 103–104, 105*f*, 130*b*
   in history as capriciousness 79, 124
   in history as constraint 47–48
   in model of deleterious recessive 25, 25*f*, 71–72, 72*f*, 85, 86*f*
   in model of heterozygote advantage 15, 24
   in model of mutation-selection balance 85, 86*f*
   in neutral theory. *See* neutral theory of molecular evolution
   in shifting-balance theory 35–36*b*
   selection mimicking 77, 79, 124
genetic variation. *See* polymorphism
genome reduction. *See* minimal gene set
Gilpin, Michael 37–38, 38*f*, 113*b*, 114 n. 3, 132
Glass, John 43–44
González-Wevar, Claudio 59–61, 60*f*, 61*f*, 61–62, 62*f*, 63
Grossi, Alexandra 57

H

Haemophilus influenzae 43–44
heterosis. *See* heterozygote advantage.
heterozygote advantage 16*f*, 18*b*, 18–20*b*, 22, 24, 36*f*, 96–97*b*, 99*b*, 102–104*b*
heterozygote disadvantage 25–26, 27*f*
heterozygosity, expected values under neutral theory 15, 17*f*, 18–19, 81, 87–90, 87*f*, 88*b*, 91–92, 95–96 n. 1, 132
historicity, Lewontin's principle of 77*f*, 77, 124
hosts and parasites. *See* coevolution of host and parasites.
Houston, Alasdair 3, 5, 34,
Hutchinson, George Evelyn 111, 111 n. 1, 111

I

Ideal Competitive-Differences Distribution (ICDD) 3–4, 32, 33*b*, 34
Ideal Free Distribution (IFD) 3–6, 34
island biogeography 10, 12*t*, 15, 81, 90–92, 94–95

J

Johnson, Kevin 9, 57, 58*f*
Justice, Lady 7*f*, 7

K

Kennedy, Martyn 5–6, 55–56
Kilsdonk, Laurens 121
Kimura, Motoo 86–88*b*, 90, 99–100, 126–128
Kingsland, Sharon 1, 8, 9, 29, 83, 90, 129
Koonin, Eugene 42–44

L

lethal mutation. *See* deleterious recessive.
Lenski, Richard 48, 49*f*
Lewontin, Richard 1, 10, 26, 68, 76–78, 89, 94–96, 95–96 n. 1, 97*b*, 97–98, 98*f*, 99–100*b*, 101–102, 124, 129
lice, bird 57, 58*f*
limit cycle 15–18, 17*f*, 73–74
limpets 58–64
Littorina saxatilis 30
Liu, Ming 26
logistic model. *See also* Lotka-Volterra model. 15, 18*b*, 18–20*b*, 21–23, 73, 74*f*, 86
long-term evolution experiment (LTEE) 48–49, 49*f*
Lotka-Volterra model. *See also* logistic model. 11, 37, 38, 68, 108–109*b*, 110*f*, 112*b*, 112–113*b*, 113–117

M

MacArthur, Robert 10, 90, 91*f*, 91, 108

MacLeod, Catriona 31–32
Maniloff, Jack 46
Markovian property 32–34*b*, 78
Marks, William (Bill) 10, 100, 100*f*, 101*f*, 101–104, 117, 125
mathematical models. *See* models, mathematical.
May, Robert 11, 73–74, 94, 108–109, 111–112
  and "devious strategies" 112, 117, 118
Maynard, Daniel 118–119*b*
Mayr, Ernst 1–2, 89
microbiome 68–69
Miles, Donald 54, 63
minimal gene set 9, 12*t*, 42–46
mite, human skin 47–48, 47*f*
models. *See also* adaptive landscape, deleterious recessive, generalized dominance, heterozygote advantage, heterozygote disadvantage, Ideal Free Distribution, Ideal Competitive Differences Distribution, mutation-selection balance, neutral theory of molecular evolution, predator-prey model, selection.
  constructionist 11, 94, 108, 125–126, 124F*f*, 130*b*
  disease 67–68, 74
  ecological 108
  food-web 94–95, 108
  logistic. *See* logistic model
  Lotka-Volterra. *See* Lotka-Volterra model
  mathematical 3, 6–7, 14*b*, 15, 28, 30, 34, 119*b*, 121
  population-genetic 94, 118 n. 5, 124
  proof-of-concept 3, 34
Moulton, Michael 30–31
Mushegian, Arcady 43–44
mutation. *See also* mutation-selection balance, neutral theory of molecular evolution. 9, 14, 48–49, 67, 84–85
  in constructionist models 10, 94–95, 99, 100–102*b*, 103–104*b*, 105, 126–128, 128*f*, 130*b*
mutation-selection balance 10–11, 12*t*, 81–83*b*, 85, 86*f*
Mycoplasma genitalium 43–44

N

Nacella 59–61, 60*f*, 62–64
natural historians 8
natural selection. *See* selection.
Nei, Masatoshi 89–91
neutral theory of molecular evolution 10, 12*t*, 15, 17*f*, 18–19, 81, 86–90*b*, 92, 95–96, 95–96 n. 1, 126–128, 130*b*, 132

  compared to island biogeography 90–92, 94–95
New Guinea, birds of 41
New Zealand, birds of 30

O

optimal foraging theory, 3–6, 8
ortolan bunting, 30–31

P

paradox of variation. *See* variation, paradox of
parameter 6–7, 10, 14*b*, 17*f*, 20*b*, 22, 30, 37–38, 67–68, 74–75, 75*f*, 75–76, 97*b*, 98, 100, 102, 112*b*, 113–119*b*, 126, 130*b*
parameter space 10, 14*b*, 97–98, 98*f*, 98–102, 105, 118, 129
parasites, coevolution with hosts. *See* coevolution of hosts and parasites
  intracellular 9, 42–43, 57
pelecaniform birds 54–56, 55*f*
phase space 14*b*, 110*f*
phenylketonuria 82, 84
phylogenetics 1–2, 47–48, 54–57, 64
pigeons 57, 58*f*
Pipek, Pavel 30–31
polymorphism 10–11, 12*t*, 94, 124, 125–128, 130*b*
population structure 26, 35, 62–63, 103
Post, Wilfred 116, 118
Powell, Jeffrey 126, 127*f*
predator-prey model *See also* Lotka-Volterra model. 17*f*, 79, 108–109
priority effects 120–121

Q

quantitative traits 68–69

R

Red Queen Hypothesis 130*b*
reed bunting, common 30–31
Rhagada 54
Roberts, Alan 112*b*, 112–113*b*, 113–115, 118
Robinson, James 115–116*b*, 118
Ruse, Michael 1–2

S

Scotland, sticklebacks in 49–51
selection 35, 76–77
  constant viability model 96–97*b*, 100
  frequency-dependent 99, 103, 125–126, 128*f*
  in a deteriorating environment 130*b*, 131*f*
Serratia symbiotica 46
Servedio, Maria 3, 6, 34
Shinohara, Naoto 118, 121
shifting-balance theory 35–37, 41

Siphonaria 61–63, 61*f*, 62*f*, 64
Smith, Gilbert 47–48
Southern Ocean 58–59, 59*f*, 61, 64
spandrels 118–119*b*, 120
Spencer, Hamish 4–6, 8, 10, 22*b*, 33*b*, 34, 56, 82, 84–85, 89, 99–103*b*, 103–104, 117, 125–128, 128*f*, 130*b*
stability. *See also* complexity and stability. 1, 18–21*b*, 22–23*b*, 46, 109*b*, 111
state space 14*b*, 110*f*
steady state 6, 10, 15, 17*f*, 81, 86, 102*b*
stickleback, three-spined 49–51, 50*f*
stochastic models, definition of 24
stochasticity. *See* chance, flavor of.
Subantarctic. *See* Southern Ocean.

T

Tay-Sachs disease 84–85
Taylor, Peter, 11, 100, 108, 112*b*, 116–118
template, flavor of 9, 12*t*, 54, 69–71, 126, 128–129
Tregonning, Ken 112*b*, 113–115
Trotter, Meredith 99, 103, 125–128, 128*f*
turnover, flavor of 10, 12*t*, 39, 81, 86–92, 102*b*, 126–130*b*, 132

V

van Tets, Gerry 54–55, 55*f*, 56
variable 14*b*, 15–18*b*, 18–20*b*, 21–26, 30, 38, 67–68, 74–75, 91–92, 111, 119*b*
variation,
  genetic. *See* polymorphism.
  paradox of 10, 95–96, 95–96 n. 1, 103
viability. *See* selection.

W

Waddington, Conrad 41, 42*f*, 44–45, 51
Weidlich, Emanuela 121
Wilson, E.O. 10, 90, 91*f*, 91–92
Wright, Sewall, 35–36*b*, 85

Y

yellowhammer, 30–32